MANAGING DOLLARS WITH SENSE

BY
MITCH SCHNEIDER

THOMSON

DELMAR LEARNING

Australia • Canada • Mexico • Singapore • Spain • United Kingdom • United States

THOMSON

DELMAR LEARNING

Automotive Service Management

Managing Dollars with Sense

Mitch Schneider

Executive Director:
Alar Elken

Executive Editor:
Sandy Clark

Automotive Product Development Manager:
Timothy Waters

Development:
Kristen Shenfield

Executive Marketing Manager:
Maura Theriault

Channel Manager:
Elizabeth Lutz

Marketing Coordinator:
Brian McGrath

Executive Production Manager:
Mary Ellen Black

Production Manager:
Larry Main

Production Editor:
Elizabeth Hough

Editorial Assistant:
Kristen Shenfield

Cover Design:
Julie Lynn Moscheo

Printed in Canada
1 2 3 4 5 XX 06 05 04 03 02

For more information contact Delmar Learning
Executive Woods
5 Maxwell Drive, PO Box 8007,
Clifton Park, NY 12065-8007
Or find us on the World Wide Web at
http://www.delmar.com

For permission to use material from the text or product, contact us by
Tel.   (800) 730-2214
Fax   (800) 730-2215
www.thomsonrights.com

Library of Congress Cataloging-in-Publication Data:

Schneider, Mitch.
Automotive service management. Managing dollars with sense/ Mitch Schneider.
   p. cm.
 ISBN 1-40182-663-6
1. Service stations--Management. 2. Business planning. 3. Budget in business. I. Title: Managing dollars with sense. II. Title.
TL153 .S34 2003
629.28'6'0681--dc21
                                        2002151917

NOTICE TO THE READER

CONTENTS

Contents

PREFACE

This shop operations guide is designed to provide a framework that helps you make consistent, high-quality, and productive service a part of your normal, everyday shop operations.

Is it everything you ever wanted to know about automotive shop management but were afraid to ask?

I'm not sure. In reality, it is a lot of what I have learned about running a successful automotive service business, both on my own and sitting at the feet of those who started this journey before I did. At the very least, it's everything you can fit into eight volumes.

The purpose of this guide is to help ensure that *great performance* is achieved every time you or one of your employees approaches a regular or potential customer to deliver automotive service. It is the only way I know to create trust and insure customer loyalty. And, great performance coupled with increased customer loyalty, trust, and operational excellence will almost always result in increased profits.

High-quality service is defined by the customer's experience while in your care and custody. Operational excellence ensures that the customer's cars will be *ready when promised* and *fixed right the first time*. Fixing the car right the first time is as much about understanding what your customer is trying to tell you when they are communicating their frustrations with the vehicle as it is about proper diagnosis and professional quality work. *Total Quality Service* is only possible through high-quality communication and knowing where you are in the process of completing a job at all times. Because the workforce is diverse, we chose to alternate between "he" and "she" from chapter to chapter to ensure gender is equally represented.

Selling service is perhaps the most difficult of all sales to make. Buying a product—something you can hold in your hand like an oil filter or a spark plug—constitutes the purchase of something you can touch or feel. Purchasing a service is something else entirely. When a vehicle owner buys a service, that person is really purchasing a promise—a promise that will be fulfilled in the future. That requires trust and a great deal of faith.

Your ability to perform is based upon a lot of things, not the least of which is technical competency. But you can't demonstrate that technical competency unless (or until) you are given the opportunity. The vehicle owner can't see technical competency. She can't touch it or feel it and you can't hang it out in front of the building like a sign.

The customer's perception of whether or not you are worthy of her trust, and whether or not you will be able to solve her automotive service problem, will be based in many cases upon any one of a number of subtle, almost subliminal, messages. These messages are broadcast and received before any one of your technicians ever picks up a screwdriver, wrench, ratchet, or socket.

Service starts when a customer sees and responds to an advertisement (an invitation to experience the quality of your service), hears about your business from a friend or relative, calls, or walks in, service includes every subsequent customer contact until the job is completed—ready when promised. Moments of truth—moments when a customer has the opportunity to come in contact with any member of your organization and form an opinion about you or your company—occur before, during, and after the job has been completed. They will ultimately define your personal success and the success of your company.

Consistent execution of company policies and procedures is the catalyst for world-class service. It is perhaps more important to be consistent at what you do than to be good at what you do. Do a great job one time, and something less than great the next, and you aren't likely to see that customer again. The more people you have working for you, the greater the challenge it is to achieve consistency. Without a comprehensive and complete set of policies and procedures, consistency is nothing more than an elusive dream.

All of this consistent high-quality service is only sustainable if your shop is operating at productivity levels that allow you to make a profit. Productivity is a by-product of your technicians' ability to execute policies and procedures flawlessly. It will take more than a visit to the tool truck or a one-day management seminar for your technicians to bill 70 percent or more of their available hours.

ABOUT THE SERIES

All eight volumes of this guide—*Total Customer Relationship Management, From Intent to Implementation, Managing Dollars With Sense, Operational Excellence, Building a Team, The High Performance Shop, Safety Communications, and Operations Management*—are designed to work together in large shops and small shops. That means that the building blocks of consistency—the policies and procedures you will find within these pages—are defined in their simplest terms because no matter how big or how small a shop is, the same kinds of tasks must be performed. The policies and procedures we define here can and will be carried out by the service writer, service manager, technician, office manager, and lot person. In many cases, the person who fills the role of service writer also serves as the service manager and perhaps even the technician. The key to success is not who does the job. The key to success is insuring that each critical task is completed accurately and consistently—the same way every time.

While this set of guides will take you through a series of step-by-step procedures, it is important to remember that delivering high-quality service is a process. To achieve success, all of the steps in the process must be performed, either consciously or unconsciously. If you are having a problem in one particular area of your business, start with the guide that addresses that need first. Next, read through the rest of the guides. See if what you find in these pages makes sense to you. *Then, read it again.* Look at your business and see what, if anything, applies to you. *Then, read it again.* Consider which of your current procedures need to remain in place. If some need to be changed, determine if some of the policies or procedures you found in these guides can be incorporated into the daily operation of your business. *Then, read it again.* Put yourself in the place of your customer. Involve everyone in your business in this exercise. You will be far more successful with their help and cooperation than you could ever be without it. Help them understand that this is how things need to be done from *this moment forward* and why. Walk through the new procedures in your shop. Pull the trigger, start the process, read the text, and begin your journey to a better quality of service, a stronger, healthier, more profitable business, and a more satisfying, less frustrating life.

ACKNOWLEDGEMENTS

Nothing of consequence occurs in a vacuum. It takes a lifetime of collaboration, cooperation, and support. The Automotive Service Management series is no exception. If it takes a village to raise a child, it takes a city to conceive, write, edit, and publish one book—let alone eight. Consequently, there are some special people I would like to thank for making the journey with me.

I'd like to thank my Delmar family: Tim Waters, Kristen Shenfield, Betsy Hough, and everyone in marketing for staying with me from conception to completion—as well as Lynanne Fowle and Megan Iverson at TIPS Technical Publishing for their meticulous attention to every detail. It is a joy to work with so many people who care.

I'd like to thank John Wick and Willi Alexander for planting the seeds, and Dick Vinet, the Duffy-Vinet Institute, Motor Service Magazine, and Champion Spark Plugs, for presenting the first shop management seminar I ever attended. It gave me confidence, but more than that—it gave me hope. And, Arthur Epstein (of blessed memory) and Ray Walker for their guidance, friendship, and patience and for recognizing in me something I could not or would not see myself.

Most of all I'd like to thank my wife, Lesley, for tolerating a lifetime of ruined dinners, working too late at the shop, and hiding out in the "Cave" writing and thinking and just generally being far away. She's not only been the biggest part of everything I've ever accomplished, believing even when I didn't—she's been the best part of everything I've ever accomplished. I'd like to thank my kids, Ryan and Dana, for sharing me with my work even when I'm sure they didn't understand. They are a blessing and an inspiration.

I'd like to thank my brothers: David for working with me for the better part of thirty-six years, and Todd for not. Both have taught me a great deal. David has taught me about patience and understanding and looking beyond my frustration to try to understand the lesson he is trying to teach me about myself. And, Todd has taught me a great deal about positive attitude, professionalism and personal courage, in addition to never saying No to anything I've ever asked of him.

Finally, I would like to acknowledge my parents—Sylvia and Jerry Schneider. They are far more than parents. They are my closest friends, my favorite teachers, and a living example of what excellence, character, commitment, dedication, sacrifice, and giving everything you have to give is all about. It's hard to relax and take it easy when your eighty-year-old parents won't, and it's just as hard to give less than your best when your eighty-year-old parents won't!

INTRODUCTION

All too often, the way most automotive repair facilities charge for their products and services has little to do with what they *must* charge in order to make a reasonable profit, and much too much to do with a combination of what the garage owner *thinks* the traffic will bear and what he or she *believes* the competition is doing. To *manage dollars with sense*, you're going to have to understand both the principles of sound business practice and the realities of the automotive service marketplace. These are both essential components of a single automotive service puzzle, and because of their reliance on one another, both are necessary to complete the picture.

We'll start by taking a look at the need for accurate and timely financial analysis and then move on to the two concepts most garage owners I have met find the most difficult and confusing to differentiate and understand: *markup* and *margin*. We'll see if we can take what we have learned and use it to effectively change our *current reality*. Then, we'll look at some critical metrics that can and will help us get our businesses on track and keep them on track—key performance indicators (KPIs) like gross profit margin (GPM) on labor, gross profit on parts, car count, average invoice, labor mix, labor content per job, service bay productivity and technician efficiency, total sales per bay, and total sales per technician.

We'll take a few minutes to focus on what seems to be the most popular method used by the majority of the shop owners across the nation to price their products and services. I call it *Oreo pricing*. It describes what happens when a shop owner or manager sets their prices in line with what he thinks the other shops in the neighborhood are charging for their products and services and then, just like the white creamy center in the middle of an Oreo cookie, sandwiches himself in between what he perceives to be the highest and lowest service providers in the marketplace.

On the surface, Oreo pricing seems like a reasonable strategy. After all, the majority of shop owners out there are paralyzed by the fear of being either at the very top or at the very bottom of the price ladder. The middle is comfortable. The middle is safe. And comfortable and safe are good. They are warm and fuzzy, and just like a glass of milk accompanied by the soft, sweet center of an Oreo cookie—warm and fuzzy is where most of us prefer to spend the majority of our time!

Finally, we'll sit down and think about what it takes to formulate a plan—a business plan: the information we have and the information we'll need. And then we'll take what we have learned and integrate it into that plan.

Managing Dollars With Sense will give you a foundation upon which to build intelligent business decisions without having to *Oreo price* your products and services, play *Follow The Leader* or resort to *Management by Hope: By Gosh or by Golly*. The rest of the books in this series will give you the information and materials necessary to build a business that will provide you and the important people close to you with a business you can depend on and a future you can look forward to.

CHAPTER 1

The Nature of Your Numbers

INTRODUCTION

This chapter introduces you to:

- The nature of business

- Where the numbers come from

- The importance of accurate and timely financial reporting

- The income or profit and loss statement

- The balance sheet

- Key business ratios

It will also identify financial data as historical data and show that it is significant only insofar as it can be used to identify areas for improvement and influence future behavior.

THE NATURE OF YOUR NUMBERS

Once upon a time, conventional wisdom suggested that if there was money in the checkbook at the end of the month, things were going pretty well. Bookkeeping and accounting were fine, but only for bookkeepers and accountants. Servicing and repairing cars was for shop owners and mechanics—people like you and me. And the world of broken cars and broken people and the world of balance sheets and income statements intersected on a monthly, quarterly, or yearly basis only.

That was then and now is now! In a world of compressed margins and competitive pressures, what was good enough for our mothers and our fathers, our brothers and our uncles, will not be good enough for the business climate you and I face today and certainly not good enough to build a better future for tomorrow. Understanding your numbers—especially the Key Performance Indicators (KPIs) that tell you at a glance just how well you are doing—is critical.

KPIs

Critical business measurements used to benchmark performance. KPIs are used to identify areas for improvement and manage progress toward the achievement of strategic goals and objectives.

Not knowing exactly where you are every day—or even every hour for that matter—with regard to margins, mark-up, profit, and loss is economic suicide. It's almost like trying to drive to New York without a map and without knowing how much gas you've got or even how much you will need.

Realistically, without understanding where the numbers come from and what they are trying so desperately to show us, we can't even begin to discuss our current financial situation—and that is what any discussion of managing dollars with sense is ultimately all about. We can't interact with an accountant or a bookkeeper unless or until we both know and understand what *they* need in order to help guide us toward financial health, security, and independence.

Consequently, it is critical to any comprehensive and substantive discussion of automotive shop management that we stop to both explore and analyze what we need to know and why we need to know it.

THE NATURE OF BUSINESS

Business is all about *raw materials* and *finished goods*. It's all about commerce—the exchange of goods and services for compensation of one kind or another.

It's about creating value, adding value, and selling the products and services we create for more

com·merce, (kom-ərs) *noun*. **1.** *Abbr.* com., comm. The buying and selling of goods and services for monetary compensation or for products and services of equal or greater value

than the cost of their creation. If my definition is correct, and I believe it is, we can start to see where the numbers come from almost from the very beginning. They are associated with the costs involved in providing those products and services. And, they are associated with the profit or the loss involved in selling those products and services.

Profit occurs if the selling price is greater than the total of the costs—loss occurs if it is not.

The problem in our world is that most automotive repair shop owners and managers can't conceive of automotive service, especially the labor side of service, as anything more than the sale of hard parts and accessories—the parts and accessories we acquire from our warehouses and jobber stores—in conjunction with the time associated with their removal and installation. More to the point, most shop owners and managers are unable to recognize the value they add to the process in terms of service, skill, ability, competence, convenience, quality, confidence, accountability, reliability, predictability, integrity and the ability to respond to customer wants, needs, and expectations.

Consequently, shop owners and managers who don't realize or understand the value of the products and services they provide the public subsidize the cost of automotive service with unrealistically low prices. Almost every problem this industry faces—the acute shortage of trained and qualified technicians, the lack of interest in automotive services as a viable profession, and the absence of succession planning or exit strategies—is the result of an inadequate revenue stream and poor compensation, both for the shop owner and for the technician.

WHERE THE NUMBERS COME FROM

The numbers come from *all* the costs and *all* the revenue associated with operating your business. They include all the costs involved in acquiring the parts and accessories we sell, as well as the costs involved in packaging them, re-selling them and then *hanging* them on a vehicle. Whether we like it not, whether we know it or not, to be successful in this or any other business environment, you and I have to know everything there is to know about every aspect of our business—and the numbers that represent that business to others is a good place to start.

The fundamental problem is that most of us refuse to take the time or make the effort to understand what the numbers are all about. Some of us don't think it's all that important, especially if we're paying someone else to do it for us. And some of us can't or won't take the time to learn how to do it for ourselves. Regardless of the reason, that lack of knowledge can devastate our ability to succeed.

That doesn't stop us from grumbling about why we have to familiarize ourselves with the numbers, or even what numbers we should familiarize ourselves with. After all, it's not like we have nothing else to do in the course of the day. In the end, though, understanding our financials is critical to managing dollars with sense. It is what our business is all about.

So what do you need to know? The first thing you need to know is that you cannot manage a shop from underneath a car. Why? Because there is no way to make a profit in today's competitive automotive service marketplace without monitoring and managing the numbers that *are* your business—and you can't do that without becoming intimately involved. You almost have to become emotionally involved with those numbers to be successful today. There's no escaping it. And you must take responsibility for your own success. Leaving it up to someone else, even a paid professional acting as an extended member of your management team, can be costly if not downright dangerous.

You don't think your business is all about numbers? You think it's about people and broken cars and service and parts and hours of operation and a hundred other different things? That may or may not be true. But, even all of those other things can be reduced to their most common denominator—the numbers that reflect the health and vitality of a business at any given moment in time or over any specific period of time.

You and I can spend as much time as necessary talking about all the various facets of a contemporary automotive service business, and we might even find ourselves agreeing on what a composite of that business should look or feel like, but if we want to explain that business to a banker, a lawyer, an accountant, or someone who might want to buy it or consider financing it, we had better be able to speak to them in a language they are comfortable with, a language they can understand. And for virtually all of the above that language is numbers.

Relating to customers is important. So is managing your employees. No one can discount the importance of instinct. Neither should intuition be disregarded. Both are an integral part of the decision making process. However, none of these will help create a successful business or strategic plan. Only the numbers that reflect what our business is all about, how it's doing, and where it's headed, will help justify the decisions we make and give them the credibility they need to help influence and impact the future

GI/GO: GARBAGE IN/GARBAGE OUT

For most shop owners and managers, each day is simply a matter of *one foot in front of the other*. Most shop owners and managers demonstrate a cavalier disregard for the need to become intimate with their numbers and even less interest when it comes to insuring that those numbers are both timely and accurate. Realistically, I can say that for many years we were far more focused on the daily tasks of *crisis management* and *damage control* than on insuring the accuracy of our financial reporting!

Unfortunately, the same rules that hold true for computer data management and programming hold true for bookkeeping and accounting. If you give your bookkeeper garbage—numbers that are unrealistic, inaccurate, or untrue—you're going to get back financial reports that reflect that same quality in their content (garbage in/garbage out).

The problem with that is simple: good numbers are critical for good decision-making and bad numbers will almost certainly lead to bad decision-making.

I suppose there is no better example than *inventory* to demonstrate my point. I'm not sure I know anyone who once a month takes the time to count every nut and bolt in the building. I'm not sure I know anyone who has not *fudged* that number at tax time to reduce income or inflate expenses. And yet, I know that everyone knows that while that number can go up and down on a daily basis, it is not elastic. It is grounded, or at least it should be, on the actual number of whatever it is you are counting being physically *on-hand*.

You would think a number that critical—a number capable of making such a dramatic difference in your bottom line—would be critical to the individual or individuals who own and operate the business, but it's not. Or, at least, it doesn't seem to be in this industry. If it were, someone would actually be counting *stuff*! And, to compound matters, I've seen the same lack of attention and concern when it comes to the other critical numbers that make up a shop's financial model.

After four generations and more than 36 years, the only thing that I can add to this discussion is the following: you can only get out of your financial reporting that which you put in. Your accountant can only advise you on the basis of the information you deliver. Everything about your business will depend upon the quality of those numbers—the value of your business when you go to sell it, your ability to seek outside financing and loans, the financial wherewithal to re-capitalize equipment, facility maintenance and training—all depend upon the story those numbers tell. Consequently, it is critical that those numbers are as timely and as accurate as you can make them.

Someone once said, "Close only counts in horseshoes and hand grenades!" Your financial reporting and the information it is based upon isn't horseshoes and it isn't hand grenades. It is the heart of your business. It is its foundation. If the numbers are both timely and accurate you can be pretty sure you will be able to depend upon what those numbers are telling you about your business. If they are *close*, their results will be close. So, the question is: Do you want success? Or do you want something *close*?

The numbers are yours. The decision to make those numbers as timely and accurate as you can make them is yours. It is, or will be, your business, so the choice will ultimately remain with you. Will you settle for *close*, or will you demand accuracy? Do you want garbage in/garbage out, or will you insist on success?

HISTORICAL DATA

There is something I mentioned earlier that I would like to revisit and emphasize here. Accounting information, bookkeeping records, and financial data in general all fall into a category that can and should be considered *historical data*. This is information that should afford us some kind of insight into something that has already occurred in the past. If you think about it, that information, that insight, is only useful insofar as it can be made relevant when projected into the future. What that means to you and me is this: having your accountant or bookkeeper merely regurgitate the numbers you furnish them back to you in a different form or format is useless—meaningless! Regardless of the form or format, you already have those numbers.

These numbers, your numbers, are only useful if they are used as a means to an end—a catalyst to change your behavior, your policies, and your procedures, in order to change your business and/or financial performance in the future. If your accountant or bookkeeper is merely picking up your numbers, moving them around on a spreadsheet and then giving them back to you without any kind of substantive analysis and suggestions to improve future performance, it's probably time to start looking for someone else who can and will!

TIME IS OF THE ESSENCE: THE NEED FOR TIMELY REPORTING

I know a shop owner whose shop was not automated until very recently. He did all his financial reporting on paper and by hand. He provided his bookkeeper with critical information about his business weekly, and she provided him with the formal financial statements that resulted from that information by the middle of the following week.

Sound like overkill? Perhaps—but it would be hard to argue that there were few people who knew more about their business and where it was financially than he did. He monitored his KPIs daily and his overall business performance by the week.

By the time most other garage owners were receiving their financials from an accounting period that had *ended* four, six, or even eight weeks earlier, he had already long since compensated and adjusted for whatever problems he had discovered... whatever he had been doing wrong... whatever he *could* be doing better!

A hiccup in the economy that made news a month after it was recognized by anyone else was old news to him. He could glance at his financials and tell you within hours what had happened and exactly when it had occurred.

I know of other shop owners and managers who receive their financial statements quarterly—weeks after the quarter has ended. Their idea of financial analysis is opening the envelope, pulling the documents out, determining if the very bottom line number was in brackets or not, and then placing those documents in the bottom right-hand desk drawer!

Getting your numbers back from your accountant a month or two after the accounting period has ended is about as relevant as someone telling you what the temperature was in your home town on a Tuesday two months ago! The information is useless: you already knew what the temperature was then and it no longer makes any difference because it is just too late to do anything about it—too late to use that information effectively or intelligently.

The bottom line is simple. Because financial data is historical by nature, time is of the essence when it comes to gathering it, analyzing it, and then drawing any kind of substantive benefit from the analysis. My recommendation here is to monitor your financial performance closely by watching your KPIs daily, weekly, and monthly, depending upon which KPIs you are looking at. If you do that, everything else will take care of itself.

If the KPIs are where they are supposed to be, the business will be performing the way it is supposed to, and the resulting financial information will be consistent with that kind of performance.

It isn't really very complicated. You and I have to stay on top of the data day-to-day. We have to insure it is complete at the end of each month and we have to see that is in the hands of the person responsible for making sense of those numbers as soon as is humanly possible. It then becomes *their* responsibility to put those numbers in a format that makes sense to them or to anyone else who might need to see them. After that, it is a matter of explaining what those numbers mean in terms you and I can understand, as well as to provide the analysis necessary to suggest intelligent alternatives for improved financial performance in the future.

SUMMARY

The key components that make it possible for a passenger car or truck to start, run, and stop effectively can be measured using a number of different tools. These tools will capture voltage readings, amperage readings, and pressures, as well as the emissions readings at the tail pipe. These measurements can then be compared to a table of known good values and a comprehensive diagnosis can be made based upon where the readings are when compared to where they should be.

What must be done to diagnose a business is really not very different. Measurements are taken in terms of the financial reporting that you as a business person are required to do, and those measurements are then compared to a table of known good values in order to determine just how you are doing.

Understanding where the numbers come from and what they are trying to tell us is critical to our ability to succeed.

Take a moment to think about the financial reports that you have available at the present time. Are they comprehensive enough? Are they providing you or your bookkeeper or accountant what you need to know? Where do the numbers come from? Are they accurate? Are they timely enough? Do they help you plan for the future or do they merely represent something that occurred in the distant past?

Think about how much time you actually spend analyzing what they are trying to tell you. Determine whether or not your financial professional is helping you understand these numbers any more clearly than you do right now. Write your questions in the chart on the next page along with any questions you might have for your bookkeeper or accountant the next time you are together.

Thoughts

Actions

Results

CHAPTER 2

Understanding Your Financials

INTRODUCTION

This chapter introduces you to:

- Profit and loss statements

- Balance sheets

- Generally accepted accounting practices

- Cost of goods sold

- Direct labor

- Expenses

- General & administrative expenses

- Gross and net profit

- Margin and markup

- Labor mix

WHAT FINANCIALS DO WE NEED TO UNDERSTAND?

There are really only two financial statements most of us ever pay any attention to-the *balance sheet* and the *profit-and-loss statement*, which includes a *statement of income* and a *statement of expenses*.

Almost all the information you really need to become a better owner or manager is available to you in these two very basic business documents. The only problems most of us have are a serious lack of business education and training, and a basic unwillingness or inability to take the time required to do more than just glance at the top line (total sales), and the bottom line (net profit or net loss), and either nod and smile, or grimace and weep.

To help you with those problems, we're going to spend a few minutes exploring a set of numbers typical of those you might find in an independent automotive service facility. To do that, however, we will first have to mentally construct such a facility. In my seminars and presentations, I refer to this hypothetical garage as Average Automotive, Inc. But that's not what we're going to call our typical garage here. We're going to call it Above Average Automotive because this shop is seriously committed to continous improvement, and because our model, our business, our garage *won't* be average after we look at the numbers, analyze them, and then use them as a tool to improve our performance!

We are going to cut our way through all the jargon—all the unfamiliar formats and principles—until we are finally able to understand just how to translate parts, labor, sublet, all our expenses, and all the people we come in contact with (employees and clients alike) into numbers. After that, we are going to learn how to massage those numbers into something meaningful—something that will tell us what we need to know. And then we are going to take the whole process one step further and learn not only how to understand that language, but how to speak it as well!

Generally Accepted Accounting Principles

In any discussion of accounting, we should take a moment to talk a little about Generally Accepted Accounting Principles (GAAP). Most accountants make a big deal about GAAP because it is the standard by which all accountancy work is measured. It is the one language all accountants both speak and understand—the universal translator of business and business numbers.

GAAP insures that, while certain elements of your financial reporting might be unique in some way, the means by which those elements are calculated and reported will be more or less the same.

In other words, your accountant would be able to determine what my accountant was trying to accomplish, because the standard used by both would be GAAP. Consequently, all income and expense statements—all profit and loss statements—would basically have the same or similar content and format.

The Income (or Profit and Loss) Statement

The profit and loss statement (P&L) is a reflection of your business and its operation over a fixed and predetermined period of time. A P&L can also be referred to as an income statement, a statement of operations, or a revenue and expense statement. They all mean the same thing. It is more or less a recording—a compilation of the results—of doing business for that fixed period of time.

It begins with a *heading*, which is the information at the top of the document that identifies the business entity and designates the time period.

The heading for our P&L statement for the month ending July 31 of the accounting period we will be using looks something like this:

TABLE 2-1 P&L STATEMENT HEADING FOR ABOVE AVERAGE AUTOMOTIVE

ABOVE AVERAGE AUTOMOTIVE REPAIR, INC. STATEMENT OF INCOME OR LOSS FOR THE MONTH ENDED July 31, YEAR

The P&L also includes a number of sections in which the heart of the financial data is reported.

The Sales section of the income statement is found just below the heading. This section reports the total sales of both the products and the services that generate revenue for your business. What kind of sales should be reported here? Certainly the three major categories of automotive sales: parts, labor, and sublet should be reported, as shown in Table 2-2. However, if you have a sales category that is unique to your business and that you feel warrants closer analysis, there is nothing to stop you from adding a category like tire sales, transmission rebuilding, or detailing, for example.

Sales

This is what the first section on the income statement looks like:

Table 2-2 SALES SECTION OF ABOVE AVERAGE AUTOMOTIVE'S P&L

Sales	1 Month	Sales (%)
Parts	$18,387	45.18
Outside Labor	$1,313	3.23
Labor	$21,134	51.93
Discounts & Refunds	($176)	−0.43
Miscellaneous Income	$37	0.09
Total Sales	$40,695	100.00

It is critical that you report your sales figures in a format that makes sense both to you and to your accountant or bookkeeper. It will take that kind of detail and accuracy to provide the kind of guidance you will need to move your business forward.

If you were to think about this in terms of your own experience, accounting information is no different from the diagnostic information you gather when trying to solve a difficult drivability problem. The more information you acquire and the more quickly it can be made available for analysis, the more quickly you can diagnose the failure, suggest the proper course of action, and repair the problem. The same rule holds true in business. The more timely the information and the greater the accuracy, the faster you will be able to use that information to fix your business.

Even if your business is not *broken*, it could probably use just a little fine-tuning!

Back to the income statement. Notice that we are reporting the sales of sublet or subcontracted products and/or services (outside radiator service, out-of-house transmission rebuilding, welding, etc.), as well as the sales of the parts we have on-site in our own inventory and those we purchase for resale. The sale of sublet or subcontracted work is really no different from the sale of any other parts or accessories we purchase that *passes through* our business. It should carry with it a margin or markup that reflects the costs involved in handling that sublet as well as the exposure to liability we accept by honoring its warranty. Consequently, the proper margins here are just as important as the proper margins on the sale of labor and parts.

For the purpose of analysis, both you and your accountant or bookkeeper must know and understand what these numbers mean in relation to your total sales. The percentage of labor sales to total sales compared with the percentage of parts sales to total sales—your *labor mix*—is one of your critical KPIs. It is an indication of how carefully you are tracking the *actual* sale of the labor hours you have *available* for sale.

Let me take a moment to explain. When I first entered this industry, the ratio of labor sales and parts sales to total sales was targeted at 50 percent each, or 50/50. In other words, every dollar of labor sales should have been accompanied by a dollar's worth of parts sales. That ratio worked in an era in which the greatest portion of your time and energy was invested in removing and replacing the broken part. That is no longer the case today. In a world of high-tech electronic engines, fuel management, suspensions, comfort, and driveline controls, the greatest investment of time and energy is often the time dedicated for research, inspection, testing, analysis, evaluation, and diagnosis. What that means in plain terms is that a 50/50 parts/labor split is no longer the goal because it is no longer effective. To be successful in today's automotive service world you almost have to achieve a 60/40 labor mix—that means that $6.00 out of every $10.00 in total sales should come from the sale of labor (or time). The closer you get to that magic 60/40 split, the closer you will be to success. Moving the labor content needle off center will mean that you are *recovering* (a word that will become increasingly more and more important as we move through the rest of this series) countless labor dollars that are currently evaporating from your service bays and are subsequently lost forever.

The first step on this journey relates back to what we were saying about moving from a 50/50 labor/parts model toward a 60/40 labor/parts model. In my own personal experience, the moment our percentage of labor sales to total sales moved above the 50/50 mix I had been taught to target as *preferred performance*, we began to see our profits increase! In fact, the higher the percentage of labor sales to total sales, the more profitable we became.

If you are looking for a reason for this dynamic, you need look no further than two critical factors. First, it results from charging (or charging more) for the time you spend on each vehicle—the time required to do a thorough and responsible job—and charging (or charging more) for diagnostic time—the time required for sufficient inspection and testing. And second, it results from the fact that your margin of profit on the labor you sell is substantially higher than your margin of profit on parts!

We haven't traveled very far, and yet we have already discovered something important, something that can make a profound difference in our ability to generate a profit.

What else can we learn from the sales portion of the income statement? We *could* make some pretty accurate forecasts about future performance if we had more detailed information of our performance in the past. With that as an example, a sales summary detailing the individual transactions and their respective categories would help us understand *how much* of each item was selling and *when*.

Here is an example of what you can do when you develop a better grasp of the numbers: If your parts sales were consistently in the 45 percent range and your labor sales were consistently in the 52 to 53 percent range, your sublet sales would continue to be approximately three percent of total sales. If you aggressively promoted your company's value-added services, as well as the quality of your technicians, products, and services to the point that you increased your gross or total sales by five percent, you could fill in the sales section of a *pro forma* (a pro forma is a statistically accurate projection or model based upon your current numbers and your intended growth) income statement because your numbers should remain relatively consistent.

If you look at Table 2-3, you can see what the sales section of our P&L statement looks like without the five percent increase.

TABLE 2-3   SALES WITHOUT FIVE PERCENT INCREASE IN PARTS, LABOR, AND SUBLET

Sales	I Month	Sales (%)
Parts	$18,387	45.18
Outside Labor	$1,313	3.23
Labor	$21,134	51.93
Discounts & Refunds	($176)	−0.43
Miscellaneous Income	$37	0.09
Total Sales	**$40,695**	100.00

And here is what it looks like *with* the five percent increase.

TABLE 2-4   SALES FOR FIRST MONTH AND SECOND MONTH WITH FIVE PERCENT INCREASE IN PARTS, LABOR, AND SUBLET

Sales	I Month	Sales (%)	I Month	Sales (%)
Parts	$18,387	45.18	$19,307	45.17
Outside Labor	$1,313	3.23	$1,379	3.23
Labor	$21,134	51.93	$22,191	51.92
Discounts & Refunds	($176)	-0.43	($176)	−0.41
Miscellaneous Income	$37	0.09	$38	0.09
Total Sales	$40,695	100.00	$42,739	100.00

As you will see later on, a five percent increase in total sales can mean an increase in net profit of thousands of dollars!

Cost of Goods Sold

The next section you will encounter as you move down an income statement is Cost of Goods Sold.

We know that you made sales, but we also know that you can't make sales without incurring costs. If you installed parts, you had to buy them from someone and someone else had to hang them on the vehicle. The *direct costs* associated with the sale of the parts you buy and the direct labor to install them will be reflected in the Cost of Goods Sold section of the P&L.

Obviously, you will need someplace to actually do the work, and that facility will require electricity, lights, lifts, heat, a telephone or two, and staffing, etc. All of these expenses will be reported in the General & Administrative Expenses section of the P&L and we'll look at that in a moment. For the time being, however, we will concentrate our attention on the Cost of Goods Sold section of the P&L statement.

TABLE 2-5 THE COST OF GOODS SOLD (PLEASE NOTE THAT WE ARE USING THE ORIGINAL NUMBERS WITHOUT THE FIVE PERCENT INCREASE)

COST OF GOODS SOLD		
Purchases: Parts	$13,335	17.08%
Change: Ending Inventory	$(657)	– 0.84%
Direct Labor	$13,006	16.66%
Sub-Contract	$5,619	7.20%
TOTAL COST	$31,303	40.09%

Table 2-5 reflects the fact that we had to purchase parts. It should only include parts purchased for a specific vehicle service or repair, or parts purchased for inventory—nothing else. Another cost, noted on the Direct Labor line of Table 2-5, reflects the amount of money we had to pay our technicians for their labor-the labor required to actually repair the vehicles. This section of the P&L should report the costs involved in paying for whatever sublet or subcontracted work we did during the accounting period we are looking at. We should note that our Cost of Goods Sold section also reflects any changes in our inventory, either up or down.

It should also be noted that on this particular financial statement there are no provisions to report changes in inventory that are related to any *work in progress* (WIP)—the jobs that you are working on at the very end of the accounting period that haven't been finished, yet still have incurred costs—like the time you might have paid a technician on a job that isn't finished or collected for yet, and/or the parts and sublet that might be attached to that same job.

WIP

WIP can have a profound impact on your financial statement. Perhaps this would be a good time to explore it in more detail. Let's see what can happen on the last day of the month when you find yourself in the process of installing a rebuilt engine on a customer's car and the job is three-quarters finished. You have already had the radiator boiled out. The engine has already been removed, the technician has already been paid for the labor involved in the removal, disassembly, and cleanup, and you have already written a check for the replacement engine and entered it on the invoice. If you calculate all the costs involved in all the work already completed on this job that is not finished yet, you can see it is a significant amount and it could throw your percentages off substantially. And while it's true that over a long enough reporting period these anomalies tend to equal out, they still must be understood and accounted for.

No one needs to be influenced by the wrong data—not while trying to make important financial decisions. Consequently, you must try to be as accurate as possible at the end of any accounting period in order to ensure that all the costs associated with the sales for that accounting period are accurately reported.

One of the reasons you look at financial data in chunks of more than one month is to establish a history and a sense of consistency regarding those numbers. It is not unusual for a typical automotive service facility to have as much as five to seven percent of total sales hanging over as *work in progress* at the end of any given month.

Gross Profit

The next thing we are going to focus our attention on is *gross profit*. Gross profit is what is left over after you take the total amount of your sales and then subtract the total amount of the direct costs involved in those sales.

It might look something like this:

Table 2-6 INCOME SECTION OF THE P&L STATEMENT

SALES	One Month	Sales (%)
Sales: Parts	$28,907	37.02
Sales: Sublet	$ 9,076	11.62
Sales: Labor	$41,101	52.64
Other Income	$ 2,183	2.80
Discounts & Returns	$ (3,187)	−4.08
TOTAL SALES	$78,079	100.00
COST OF GOODS SOLD		
Purchases: Parts	$13,335	17.08
Change: Ending Inventory	$ (657)	−0.84
Direct Labor	$13,006	16.66
Sub-Contract	$ 5,619	7.20
TOTAL COST OF SALES	$31,303	40.09
GROSS PROFIT	$46,777	59.91

Generally speaking, the higher the gross profit percentage, the greater the financial stability of the business.

It would be great if all that money was ours to keep, but it's not! There are general and administrative expenses that must be accounted for that will eat away at that gross profit. Aside from that, there are a lot of other people who are intent at getting their hands on your checkbook. There is an assortment of taxes—local, state, and federal—as well as a host of different kinds of insurances. And then there is the whole universe of other expenses involved in running any small business today. As owners, you and I get paid only after everyone else has been paid. It's a part of the risk we accept when we choose to become entrepreneurs. If we want anything left over for ourselves, we had better be very careful about what those expenses are. And we had better be just as careful about controlling them, not just when things are *tight* or *bad*, or when the market is down, but when they are good as well. The worst thing that could happen is that things might never get *tight* or *bad*!

General and Administrative Expenses

This is how Above Average Automotive reports their general and administrative expenses:

TABLE 2-7 GENERAL AND ADMINISTRATIVE EXPENSES

	One Month	Sales (%)
Accounting	$538	0.69
Advertising	$252	0.32
Automobile Expenses	$145	0.19
Bad Debts	$2,730	3.50
Bank Charges	$17	0.02
Credit Card Discounts	$374	0.48
Depreciation/Amortization	$627	0.80
Donations	$–	0.00
Dues/Subscriptions	$216	0.28
Entertainment & Promotion	$190	0.24
Equipment Lease & Rentals	$482	0.62
Freight Out	$–	0.00
Health Insurance	$1,499	1.92
Insurance: General	$1,815	2.32
Insurance: Workers Compensation	$601	0.77

TABLE 2-7 GENERAL AND ADMINISTRATIVE EXPENSES (CONTINUED)

	One Month	Sales (%)
Interest	$2,868	3.67
Laundry & Uniforms	$508	0.65
Legal	$–	0.00
Meals & Entertainment	$974	1.25
Office Supplies	$367	0.47
Professional Fees	$2,000	2.56
Rent	$4,500	5.76
Repairs & Maintenance	$829	1.06
Salaries: Indirect Labor	$1,544	1.98
Salaries: Office	$414	0.53
Salaries: Officers	$3,330	4.26
Security & Alarms	$26	0.03
Shop Supplies	$366	0.47
Small Tools	$586	0.75
Taxes: City	$263	0.34
Taxes: DMV		0.00
Taxes: Payroll	$2,132	2.73
Taxes: Property	$1,317	1.69
Telephone & TWX	$422	0.54
Training	$175	0.22
Utilities	$581	0.74
Waste: Hazardous Materials	$290	0.37
TOTAL OPERATING EXPENSES	**$32,976**	**42.23**

As you can see, the normal, everyday expenses generally associated with running a contemporary automotive service business can easily become overwhelming and erode gross profit numbers very quickly. Consequently, it is critical that the percentage of gross profit is high enough to sustain those expenses. Otherwise the result is that you are working for everyone else's benefit and not your own!

If your gross profit is not adequate, you are strangling your business. You are asking too great a sacrifice of your family, yourself, and quite possibly all those who work for you; and you are very likely subsidizing the repairs of every one of your customer's vehicles.

Your customers may be wonderful, but not wonderful enough for so severe a sacrifice!

Net Profit

There is still one more area to focus on before we leave the P&L statement. We know what our total sales were, and we know what our cost of goods sold was. We subsequently know what our gross profit was and because we have an accounting of our general and administrative expenses, we know what our net income before taxes should be. But if we truly want to know what our bottom, bottom line really is, we have to complete the calculations that will allow us to really know what we are left with after everyone else gets *paid* and that means subtracting the estimated tax burden identified in Provision For Taxes in Table 2-8.

TABLE 2-8   NET INCOME OR LOSS AFTER TAXES

GROSS PROFIT	$46,777	59.91%	Gross Profit
General & Administrative Expenses			
(Schedule Attached)			
Total General & Administrative Expenses	$32,976	42.23%	
Other Income & Expenses			
Interest Income	$1,828	2.34%	
Miscellaneous Income	$1,024	1.31%	
Total Other Income & Expenses	$2,852	3.65%	
Net Income or Loss	$16,652	21.33%	Net Income or Loss before Taxes
Provision for Taxes	$1,810	2.32%	
Net Income or Loss	$14,842	19.01%	Net Income or Loss after Taxes

THE BALANCE SHEET

The *balance sheet* is a financial document that reflects the assets, liabilities, and capital (owner's or stockholders' equity) of a business at a specific moment in time. A balance sheet created at the close of business today will not be the same as the balance sheet created for the same business yesterday, nor will it be that same as the one we would create for that business tomorrow.

It is called a balance sheet because it must, upon its completion, be in balance reflecting the following accounting equation:

Assets - Liabilities = Capital

What that means is that what you *own* (the assets of the business) minus what you owe (the liabilities of the business) equals what is left over (the capital or owner's equity—or stockholders' equity if it is a corporation).

While the balance sheet reflects a precise moment in time, it is cumulative in nature in that it reflects everything that has occurred in that business since its formation. And although a balance sheet could be created at any moment in a business's history, it is generally associated with the end of an accounting period. That means if you receive your P&L statement for the month ended July 31 of a certain year, it will usually be accompanied by a balance sheet calculated for the close of business as of July 31 of that year. The reason is relatively simple to understand—the balance sheet requires information that is available only after the creation of a P&L statement and its accompanying income and expense statements, and those documents are generally tied to the calendar in terms of End of Month, End of Quarter, and End of Year (see Table 2-9).

Assets

Assets are anything that is owned by a business. Cash is the most obvious asset. However, there are many other assets a business can hold in addition to its cash reserves or receipts. Assets are classified as either current, fixed, or other assets.

Current assets are those assets which can easily be converted into cash within one year, or that will be consumed within that same period. Again, cash is the most obvious current asset. However there are many others. Your accounts receivable, money that is owed to you by your clients, is certainly another. And, hopefully, your receivables meet the criteria of this definition and will convert to cash with 12 months. Inventory is another asset that can generally be converted into cash easily within one year as well. Office supplies, prepaid expenses, and a host of other assets can be considered current as long as they meet the same criteria.

There is another category of assets that are identified as fixed. These are assets that are not likely to be converted into cash by being sold or consumed within one year, but they are still owned and utilized by the business. A good example of a fixed asset would be any real estate the business owns—any vehicles, buildings, machinery, or equipment (particularly large equipment), all of which will be listed on the balance sheet at their original cost.

Another example of a fixed asset would be leasehold improvements such as improvements made to the physical plant, the permanent installation of lifts or other capital equipment, remodeling, or refurbishing.

Investments are also considered assets, and they are classified short- or long-term depending upon when they mature.

On a typical balance sheet, assets are listed in the order of their liquidity—how easily they might be converted into cash—with current assets listed first.

Liabilities

Liabilities are considered to be *claims against* the assets of a business, both by creditors and by the owners of the business. Just as there are current and fixed assets, so there are current liabilities and long-term liabilities: claims that are immediate, due and payable within one year; and claims that are due further off in the future.

Current liabilities are generally satisfied by current assets. Examples of current liabilities are accounts payable, money that you owe your suppliers; taxes payable, money that you owe the government; salaries payable, money that you owe your employees; and notes payable (if the obligation is due within one year), money that you owe someone else.

TABLE 2-9 THE BALANCE SHEET

ASSETS

	November 30, 2000		November 30, 1999	
CURRENT ASSETS				
Cash	$52,039.73		$66,647.83	
Accounts Receivable	3,817.84		8,586.05	
Suspense			847.26	
Inventory	15,801.92		16,254.55	
Prepaid Expenses	419.70		1,227.84	
TOTAL CURRENT ASSETS		72,079.19		93,563.53
FIXED ASSETS				
Automobiles	$40,401.48		$40,401.48	
Furniture & Fixtures	6,929.30		6,929.30	
Shop Equipment	138,260.12		130,252.05	
Tools	52,250.09		52,250.09	
Office Equipment	17,588.80		17,588.80	
Leasehold Improvements	5,556.30		5,556.30	
Accumulated Depreciation	(245,158.01)	15,828.08	(226,168.59)	26,809.43
TOTAL FIXED ASSETS		15,828.08		26,809.43
OTHER ASSETS				
Due from Stockholders	15,108.66		12,608.66	
TOTAL OTHER ASSETS		15,108.66		12,608.66
TOTAL ASSETS		$103,015.93		$132,981.62

LIABILITIES AND STOCKHOLDERS EQUITY

	November 30, 2000		November 30, 1999	
CURRENT LIABILITIES				
Equipment Loans	$821.00		$2,000.00	
Loan from/to Schneider Ptnshp	2,865.72		2,865.72	
Accounts Payable	9,161.68		13,726.08	
Customer Deposits	2,477.30		4,477.30	
Sales Tax Payable	1,565.53		1,751.49	
Accrued Rent			1,000.00	
Accrued Expenses	2,811.50			
Corp. Tax Payable-Federal	2,586.00		6,252.00	
Corp. Tax Payable-State	1,554.00		3,845.00	
Note Payable-GMAC	5,437.09		5,437.09	
Loan Payable-Camarillo Bank	4,026.17		4,026.17	
TOTAL CURRENT LIABILITIES		33,305.99		43,380.95
LONG-TERM LIABILITIES				
Note-GMAC	$40,306.78		$45,408.77	
Loan Payable-Camarillo Bank	564.81		5,585.49	
TOTAL LONG-TERM LIABILITIES		40,871.59		50,994.26
TOTAL LIABILITIES		74,177.58		96,375.21
STOCKHOLDER'S EQUITY				
Capital Stock	1,000.00		1,000.00	
Retained Earnings	27,838.35		35,606.41	
TOTAL EQUITY		28,838.35		36,606.41
TOTAL LIABILITIES AND STOCKHOLDERS EQUITY		$103,015.93		$132,981.62

Longterm liabilities, usually due and payable far in the future (more than a year), could be mortgages payable or notes payable (the long-term portion of a note).

Capital or Net Worth

Capital and net worth are synonyms—two ways of saying the same thing. They are a means of recognizing the owner's claim on the assets of a business. They could be comprised of the capital accounts for the proprietor or the partners reflecting their original investment plus profits reinvested in the business—such as capital stock, with a value assigned at the time the stock was issued. Or they could reflect the accumulated or retained earnings of the business-those profits reinvested in the business after paying some kind of dividend.

For reporting purposes, long term liabilities will follow current liabilities just as fixed assets follow current assets.

What does a balance sheet look like? It can generally look one of two ways-it can be displayed in report form or in fccount form. In either case, the numbers will be the same. Only the formal presentation of the information will differ.

In the end, the two formats are just different ways of displaying the same information and which one you choose will be a matter of preference. All that really matters is what the balance sheet is trying to tell you and what you should be looking for. The bottom line will always be the bottom line! On a P&L statement you are looking for a high net profit, and on a balance sheet you are looking owner's equity that is continually growing and increasing.

Additional Financial Statements

There are two more financial reports that we should at least mention here and they are the Statement of Capital and the Statement of Cash Flow. The Statement of Capital is primarily a report documenting what has happened to the owner's capital—his or her *stake in* the business, during the year. Has it increased? Has it decreased? Has it remained the same? Was there any reinvestment of capital back into the business or were the profits all taken out in draws? Were there losses that diminished the total capital?

The Statement of Cash Flow, which was at one time referred to by accountants and bookkeepers as the Statement of *Funds*, describes both the sources and applications of the cash that flows through the business.

The P&L statement is calculated first, the Statement of Capital next, and the balance sheet, which requires data from the first two statements to be completed, is calculated last.

RATIOS AND ANALYSIS

Now that we have the numbers, what do we do with them? How do we analyze them and use them to increase our profits and improve our current reality?

As we have mentioned earlier, your inventory is one of the most critical areas you can focus your attention on in business because of the profound and dramatic impact it can have on profits. The most common way to *track* inventory is by watching inventory *turns*. Inventory turns or turnover refers to how fast those parts and accessories you have on the shelf in the back room are being removed for installation, then repurchased and returned to the shelf to be sold again.

It is calculated by dividing your cost of goods sold per year (at cost) by your average inventory per month. The equation looks like this:

Cost of Goods Sold Per Year ÷ Average Inventory = Inventory Turnover

This is a critical calculation because one of the most important characteristics of any successful business is cash flow—something we will take a look at in a minute. Money that is tied up in inventory, that is not moving parts and accessories off the shelf, that does not flow, is stagnant. It is not a return on investment for the capital tied up in that inventory. Money that is tied up in inventory that is not turning cannot be used for anything else. Calculating inventory turns will help you to determine whether or not that asset—that cash—can be put to better use in some other way, through some other investment, than it will gathering dust on a shelf.

How Does it Work?

Let's say Above Average Automotive had a Cost of Goods Sold at the end of the last accounting period of $288,310 and the average monthly inventory was $13,300. Using our formula, the math looks something like this:

$288,310 ÷ $13,300 = 21.67

According to our formula, this inventory isn't turning... it's spinning! But is that *good*? That depends upon a number of variables, such as whether or not *all* the items in inventory are turning at the same rate. It is possible to have your entire inventory turning over an almost equal number of times, or you could find only half of your inventory turning at twice that rate while the other half isn't moving at all!

You may find yourself or your technicians waiting for parts to be *hot shot* delivered because the parts you need aren't on the shelf, and that these delays are seriously impacting your shop's productivity in a negative way. You could *invest* more dollars in inventory, and possibly experience an increase in shop productivity that could return a higher percentage on your investment than you could get anyplace else. Or you could actually decrease your on-shelf inventory, allowing yourself to become increasingly more and more dependent upon your suppliers.

The question here is not whether this ratio is a good ratio or not, it is whether or not you have the information you need to make an intelligent decision. Or will you make those difficult choices with nothing to go on but instinct and intuition?

The important thing to realize is that you can affect some degree of control over your numbers and that you don't have to be controlled by them. But if the first step toward freedom is education, the second has got to be understanding. You can't affect inventory turns until you first understand what they are and how they impact your business. You won't be able to do that until you understand the difference between a net profit on paper and cash in the bank!

There are basically three different kinds of ratios that we can look at. We can look at ratios that measure how liquid the assets of a company are (liquidity ratios), how quickly we can turn the assets of a company into cash. We can look at financial performance ratios, how well we are able to control our costs or pricing policies, our expenses when compared to our industry or other industries, and our earning power. And we can look at certain asset performance ratios.

Inventory turnovers is an asset performance ratio. So is gross margin return on investment, and sales per employee, etc.

Liquidity Ratios

Let's take a look at the liquidity ratios first. The most commonly used of these ratios are the *current ratio* and the *quick ratio*.

The quick ratio is calculated by the following equation:

Quick Assets ÷ Current Liabilities = Quick Ratio

What are quick assets? That depends on who you ask! I found it calculated three different ways in three different textbooks. However, all the formulas always included cash and accounts receivable.

Why is this ratio so important and who is it likely to be important to? Well, try getting past your banker without having the necessary information to calculate a quick ratio and see how easy it is to get that loan you so desperately need! And, since a quick ratio represents a measure of just how flexible your company might be, how able it will be to meet a significant financial challenge, it is very important.

How does Above Average Automotive stack up? Above Average had $53,039.73 in cash and $3,817.84 in accounts receivable. Quick assets therefore totaled $55,857.57 and current liabilities were $33,857.57.

TABLE 2-10   QUICK RATIO CALCULATIONS

Cash	$52,039.73
Accounts Receivable	$3,817.84
Quick Assets	$55,857.57
Current Liabilities	$33,305.99

Dividing quick assets by current liabilities leaves you with a quick ratio of 1.68:1. A quick ratio of 1.5:1 or 1:1 is considered good or at least adequate for most industries, according to my accountant. That means that Above Average Automotive is *above average*!

The current ratio is another liquidity ratio that we should be aware of. It is not generally considered as accurate an indicator of how liquid your business is because it contains some items, inventory as an example, that are not really all that liquid.

The formula for current ratio is:

Current Assets ÷ Current Liabilities = Current Ratio

Current ratio is, however, very important to the people who really study your financials-primarily your banker or someone asked to provide your company with credit, because it is an indication of a business's short term ability to repay a debt.

Above Average has a current ratio that looks like this:

TABLE 2-11 CURRENT RATIO

Current Assets	$72,079.19
Current Liabilities	$33,305.99
Current Ratio	2.16

Expressed as a ratio, the result is stated as 2.16:1 (two point one six to one) or about 2:1.

You don't have to be a Certified Public Accountant (CPA) to realize that if the current ratio measures short term debt-paying ability, the higher the ratio of current assets to current liabilities the better. One of the resources I used for this research suggests the guideline should be at least 2:1 (two to one).

The final liquidity ratio to look at is the ratio of accounts receivable to accounts payable. Expressed as a formula, it looks like this:

Accounts Receivable ÷ Accounts Payable = Receivable to Payable Ratio

TABLE 2-12 ACCOUNTS RECEIVABLE TO ACCOUNTS PAYABLE RATIO

Accounts Receivable	$3,817.84	
Accounts Payable	$9,161.68	
A/R to A/P Ratio4	1.7%	or 0.42:1

Above Average Automotive has a ratio of receivables to payables of 0.42:1. Is that good or bad?

The important thing to remember here is to control your accounts receivable and to keep your payables current. Watch for trends. Are the receivables growing? Is the company able to pay its suppliers when the payables are due? Are they increasing? Is this the sign of a slow or a depressed economy? And, perhaps most important, is there anything you can do to change or mitigate any of these conditions?

Financial Performance Ratios

The next series of ratios to be concerned with are the financial performance ratios. They tell us how we are doing financially, how profitable we are.

Gross Profit Margin (GPM) Ratio

The first, and certainly one of the most important, is the GPM ratio. It is calculated by dividing gross profit dollars by net sales dollars.

Gross Profit in Dollars ÷ Sales in Dollars = Gross Profit Margin Ratio (or percentage)

This ratio is a critical measurement of your company's ability to meet all of its financial obligations. It also indicates your performance as a manager—your ability to control expenses, affect sales, and maintain effective profit margins.

Gross profit percentage is one of those numbers anyone and everyone who ever looks at your business will want to know, and it is certainly an important one for you to monitor. Check it often and insure that you are constantly trying to improve it!

GPM ratio isn't quite as popular or well known outside most accounting circles, but it is equally as important.

TABLE 2-13   GPM RATIO

Gross Profit Margin Ratio		
Gross Profit	$46,777	
Sales	$79,084	
GPM Ratio	59.15%	or 0.6:1

For most businesses, a GPM of 50 percent or better is excellent. However, there are a number of people in our industry that suggest that 60 percent should be the ideal.

Most people will tell you there are only three ways to move that ratio in a positive direction. You can increase sales. You can cut or reduce your expenses. Or you can move the needle by buying at a deeper discount, or, better, by raising your prices or by taking all the discounts available. For instance, a 2 percent discount offered for account payment by the 10th of the month will add up to a 24 percent savings on the year off an average month's payable when the amount of the discount is accumulated over a full calendar year. That can represent a substantial amount of money. A case could even be made for borrowing the money to pay your bills by the discount date. The amount of interest on any reasonable loan would still be far less than the 24 percent savings. So, take all available discounts!

There is one other way to control expenses that we will discuss later on in this set of guides and that is by *recovering* all or some of a number of different expenses.

Increasing sales, decreasing expenses, or moving your gross profit margin in a positive direction by even a small amount can have an instant and positive impact on your bottom line. That's what managing dollars with sense is all about

Let's *test drive* the numbers to see what happens when we increase sales by about 10 percent and look for better prices on our purchases, increasing our margins by a few percentage points. We're going to charge more for diagnostic time and increase our labor sales while we trim a few dollars off our expenses.

TABLE 2-14 ENHANCED P&L STATEMENT

Above Average Automotive, Inc.				
STATEMENT OF INCOME FOR THE MONTH ENDED APRIL 30, 1999				
	One Month	%	One Month	%
SALES				
Sales: Parts	$27,530	37.05	$30,283	37.00
Sales: Sublet	$8,644	11.63	$9,508	11.62
Sales: Labor	$39,144	52.67	$43,059	52.61
Other Income	$2,183	2.94	$2,183	2.67
Discounts & Returns	$(3,187)	−4.29	$(3,187)	−3.89
TOTAL SALES	**$74,314**	**100.00**	**$81,845**	**100.00**
COST OF GOODS SOLD				
Purchases: Parts	$13,335	17.94	$14,002	17.11
Change: Ending Inventory	$(657)	−0.88	$(657)	−0.80
Direct Labor	$13,006	17.50	$13,656	16.69
Sub-Contract	$5,619	7.56	$5,900	7.21
TOTAL COST OF SALES	**$31,303**	**42.12**	**$32,901**	**40.20**
GROSS PROFIT	**$43,011**	**57.88**	**$48,945**	**59.80**
General & Administrative Expenses (Schedule Attached)				
Total General & Administrative Expenses	$29,719	39.99	$30,574	37.36
Other Income & Expense				
Interest Income	$1,828	2.46	$1,828	2.23
Miscellaneous Income	$1,024	1.38	$1,024	1.25
Total Other Income & Expense	$2,852	3.84	$2,852	3.48
Net Income or Loss	$16,144	21.72	$21,223	25.93
Provision for Taxes	$1,810	2.44	$1,810	2.21
Net Income or Loss	$14,334	19.29	$19,413	23.72

TABLE 2-14 ENHANCED P&L STATEMENT (CONTINUED)

GENERAL & ADMISTRATIVE EXPENSES				
Accounting	$538	0.72	$538	0.66
Advertising	$605	0.81	$605	0.74
Automobile Expense	$145	0.20	$145	0.18
Bad Debts	$574	0.77	$574	0.70
Bank Charges	$17	0.02	$17	0.02
Credit Card Discounts	$374	0.50	$374	0.46
Depreciation/Amortization	$627	0.84	$627	0.77
Donations	$	−0.00	$	−0.00
Dues/Subscriptions	$216	0.29	$216	0.26
Entertainment & Promotion	$190	0.26	$190	0.23
Equipment Lease & Rental	$482	0.65	$482	0.59
Freight Out	$	−0.00	$	−0.00
Health Insurance	$1,899	2.56	$1,804	2.20
Insurance: General	$1,815	2.44	$1,815	2.22
Insurance: Workers Comp.	$601	0.81	$601	0.73
Interest	$2,868	3.86	$2,868	3.50
Laundry & Uniforms	$508	0.68	$508	0.62
Legal	$	−0.00	$	−0.00
Meals & Entertainment	$974	1.31	$974	1.19
Office Supplies	$367	0.49	$367	0.45
Professional Fees	$175	0.24	$175	0.21
Rent	$4,500	6.06	$4,500	5.50
Repairs & Maintanence	$829	1.12	$829	1.01
Salaries: Indirect Labor	$1,544	2.08	$1,544	1.89
Salaries: Office	$414	0.56	$414	0.51
Salaries: Officers	$3,300	4.44	$4,300	5.25
Security & Alarm	$26	0.03	$26	0.03
Shop Supplies	$366	0.49	$366	0.45
Small Tools	$586	0.79	$586	0.72
Taxes: City	$263	0.35	$263	0.32
Taxes: DMV	$	−0.00	$	-0.00
Taxes: Payroll	$2,132	2.87	$2,132	2.61
Taxes: Property	$1,317	1.77	$1,317	1.61
Telephone & TWX	$422	0.57	$401	0.49
Training	$175	0.24	$175	0.21
Utilities	$581	0.78	$552	0.67
Waste: Hazardous Materials	$290	0.39	$290	0.35
TOTAL OPERATING EXPENSES	$29,719	39.99	$30,574	37.36
INCOME BEFORE FEDL INCOME TAX	$13,292	17.89	$18,371	22.45
Provision for Income Tax	$1,942	2.61	$1,942	2.37
NET INCOME	**$11,3501**	**5.27**	**$16,429**	**20.07**

We increased our parts, labor, and sublet sales by 10 percent through a combination of raising our prices and increasing the amount of work we are doing. It isn't reasonable to assume that our costs would remain the same if the volume of work increased (although they wouldn't go up if we just raised our prices!), so we raised the costs associated with sublet, direct labor, and cost of goods cold by only 5 percent. The fact that they did not increase as much as sales is because we increased our prices as well as our volume.

We also took a look at our chart of expenses and made a few changes there as well. We reduced one of our insurance expense categories by 5 percent because we joined a professional association and reaped the benefits that accompany being a part of a group. We changed our telephone carrier and saved another 5 percent. In the process, we decided we were a bit under-paid and voted officer's salaries a $1,000 per month raise!

All in all, we lowered three expenses and increased another. However, in the end, we watched as $5,079 of additional *profit* floated to the bottom line! Is it as easy in the real world as it is on a spreadsheet? No, it isn't. But that doesn't mean you shouldn't give it a try. The only thing you can be sure of is that if you don't do *something*, nothing will happen, except possibly for things to get worse!

Return on Equity

Return on equity is another ratio your banker will be concerned with. It measures the return the stockholders/shareholders/owner-principals will enjoy. It is ultimately a measure of just how prof-itable the company really is. The formula is:

Net Income ÷ Stockholders' Equity = Return on Equity

We're going to make this calculation using the same balance sheet numbers in Table 2-9 to iden-tify stockholder's equity. In this example, we show a net income of $8,272.35 and stockholder's equity of $59,912.32.

TABLE 2-15   RETURN ON EQUITY

Return on Equity		
Net Income	$8,272.34	
Stockholders' Equity	$59,912.32	
ROE Ratio	13.8%	or 0.14:1

There are a number of other performance ratios to both learn and master: operating profit per-centage, net income before taxes, return on assets. In fact, there isn't any single element of your business that can't be examined as a ratio of some kind when compared with any other element of your business, particularly sales.

There are a host of ratios that can be found in Appendix A at the end of this guide. However, most good accountants yearn for the opportunity to show you what they know and what they can do to help you help yourself. They're just a little odd. You see, they like numbers almost as much as you like cars. When they start to spin out of control, force them back to reality. Make them explain what they mean and what they need you to do in terms you can understand. Be a customer and give someone else the opportunity to be the professional.

A Course of Action

The key here is *not* to settle for *good enough*. Don't just glance at your financials a month or two after you receive them and then slide them into the bottom desk drawer. Don't wait until it's too late to see just how powerful this kind of knowledge can be. And, most of all, don't walk around in a stupor bemoaning the fact that you don't understand, because it really isn't anywhere near as difficult as an intermittent drivability problem!

Determine what you need to know and insist on the right information from your accountant. Make it clear that you need accurate data as quickly as he or she can get it to you. And don't just accept those numbers without making sure you understand them at least well enough to know when they don't make sense or when you think you are not getting what you need!

I have left more than one accountant when the flow of information and the appropriate guidance proved undependable. I made those hard choices after I found problems in our own financial statements that had been there for years without anyone ever realizing it. Don't find yourself in the same awkward and uncomfortable position I did when I finally screwed up enough courage to ask how something like this could happen. And don't ever remain with someone who answers you the same way that I was answered when my accountant said "Well, no one has ever asked me that before!"

Whatever you do, don't ever pay an accountant that is unwilling or unable to personally review your financials. But remember, that means you will have to review them as well!

Remember what we said about GI/GO (garbage in/garbage out)? Give your accountant erroneous and unsubstantiated numbers and you will receive financial documents that are inaccurate and undependable. Then ask yourself what will happen the next time you need to apply for a lease or a loan and you hand your banker or the leasing agent a set of financial documents that are based upon those inaccuracies.

What will happen when you go to sell your business and you can't get what it's worth or even recover your investment because you have been understating profits to control your tax liability. It is a good accountant's responsibility to see that you don't pay any more taxes than you have to. But it is also his or her responsibility to make sure that you understand that the only way the value or worth of a business can be calculated is by its earnings potential.

If the numbers aren't there, if they aren't adequate to support your demand, then the selling price, what you are asking for that business, won't be there either.

Comparisons: Month to Month, Year to Date

There are really only two significant comparisons that we should be concerned about when it comes to our financials. One is a comparison of our current performance with our own past performance and the other is our performance compared to established industry standards. Unfortunately, there really is no accepted industry standard to judge our performance by. Independent garages are small businesses and small businesses are notorious for not having the most accurate financial reporting. Aside from that there are almost as many different kinds of automotive service businesses as there are businesses. There are partnerships, sole proprietorships, corporations, and a host of others. Some owner/principals take a draw based upon what is in the cash box at the end of the day and others put money back in the till at the end of the day to help support a failing enterprise.

Consequently, the only meaningful comparison is the one you are able to make with your performance of the previous month or with your performance for the same month a year or two or three ago. The key areas of concern there will be: volume, gross profit, and net profit, and how much money you have left to reinvest in your business, your people, your equipment, and your industry.

UNDERSTANDING YOUR FINANCIALS

The lesson throughout this whole discussion should be that you are for the most part in control of your own financial destiny. They are, after all, your numbers and it is, after all, your business. Consequently, the responsibility will ultimately be yours as the owner or manager of that business enterprise. Since they *are* your numbers and it is your responsibility, you *can* take a proactive role in changing those numbers—and as a result, the bottom line.

You will be the only person in your universe with the desire or the ability to impact those numbers. You are the one everyone will be looking to for guidance. You will decide what the target is, how much to increase your sales or cut your expenses. The only real question is: How? How do you do it? Where do you focus your attention to get the best, most dramatic results?

You do it where you have the most control! You monitor the ratios and keep a careful watch on the percentages. You push the margins and manage the expenses.

The more you measure and the more you manage, the more control you will have.

You have the basic understanding you need to use the tools you have to start that journey now. Take the time. Make the time. And focus.

SUMMARY

In order to successfully manage an automotive service business today you must become intimately involved with the numbers that really represent your business and its ability to perform. These numbers are the language of the financial community, and being able to *read* and *write* these numbers is critical when the time comes to communicate with members of that community.

These numbers are communicated in GAAP—Generally Accepted Accounting Principles—and are spoken and understood by professional accountants and bookkeepers. Since our success—our ability to generate profits and reduce losses—is based upon what they are trying to tell us, it would be wise for us to understand this language as well. GAAP dictates that these numbers will be reported in a generally accepted format that is both formal and structured. The forms include the income statement, the balance statement, and all their respective components, as well as a series of critical ratios (see Appendix A at the end of this guide for a more complete list of ratios).

Because accounting information is *historical information*, the numbers must be timely and accurate for them to have impact on the future.

Think about these relationships and what the numbers are trying to tell you and write down what you will be doing to embrace these numbers in the chart on the next page. Also, take a moment to think about what kinds of questions you will be asking your bookkeeper or accountant when you get your financials back at the end of the next financial reporting period, and write them down as well.

Thoughts

Actions

Results

CHAPTER 3

Margin & Markup—Inventory, Parts and Accessories

INTRODUCTION

This chapter introduces you to the business principles of:

- Margin

- Markup

It will then discuss their application, including how to benchmark adequate percentages of each.

It will also look at the relationship that exists between manufacturing, distribution, and the service industry in regard to inventory.

- Inventory

- Hotshot delivery

- Captive or dealership parts

MARGIN AND MARKUP

In all the years I have been writing and presenting automotive shop management seminars to this industry, I have found no single concept that confuses automotive repair shop owners and managers more than margin and markup. The two are used synonymously, when in fact they couldn't be more different. I have heard shop owners apply a fairly accurate definition of the principle of margin to markup and vice versa without a second thought. But more than that, I have watched the majority of the shop owners and managers I have encountered completely shut down the minute the subject of margin and/or markup was raised in discussion.

In all honesty, this was perhaps the simplest, and at the same time, the most difficult concept for me to master personally, and I would be lying to you if I told you that it didn't still present a problem for me from time to time. I believe margin and markup are problematical for most of us because they both deal with the same thing: the difference between the *selling price* of an item and its *cost*. The problem is inherent in the *way* they deal with that difference.

There are many definitions that are appropriate for our discussion of selling prices, costs, and their differences. The one below is more *classical* in a way and certainly more consistent with the kind of definition you might find in a textbook. For our purposes, we will be defining margin as the difference between the *cost* and the *selling price* of an item related to that *selling price*. This relationship is expressed in the definition of margin below—highlighted in gray.

mar·gin, (mär'-jin) *noun. Abbr.* marg. reward, return, profitable return, profit, margin of profit, bottom line, gain *Synonymous* with *discount* 1. A measure, quantity, or degree of difference. 2. Economics. **a.** The minimum return that an enterprise may earn and still pay for itself. **b. The difference between the cost and the selling price of an item for sale.**

Markup is similar, but different. It can be defined as the difference between the *cost* and the *selling price* of an item as it relates to the cost of that item. This relationship is expressed and highlighted in gray in the definition of markup below.

mark·up, (märk'-ŭp) *noun.* 1. A raise in the price of an item for sale. **2. An amount added to cost in calculating a selling price, especially an amount that takes into account overhead and profit.**

Because markup is calculated based upon the cost of an item and margin on the selling price, the percentages of margin and markup cannot be the same. In other words, if you were to calculate a 50 percent margin of profit and a 50 percent markup for the same item, they would not equal the same dollar amount.

As we have already seen, the difference between the selling price of an item and its cost can be defined in many ways. Margin and markup are only two of them. What both definitions have in common is that the difference between the *cost* and the *selling price* they refer to is usually called *profit*. This difference, or profit, is what margin and markup are all about.

The confusion occurs because the only known or given amount you generally have with regard to pricing is the cost of the item you have purchased for resale. The end result, the desired goal, the correct selling price is generally unknown until calculated. Even in those cases where the manufacturer or parts supplier provides a Manufacturer's Suggested Retail selling price (MSRP), that MSRP is a suggestion and nothing more.

If that MSRP is inadequate, if it is below the minimum margin you have established as a baseline for survival in your business, it is your responsibility as the owner and/or manager of the facility to insure that your margin is adequate by increasing it appropriately!

prof·it, (prŏf'-it) *noun.* **1.** An advantageous gain or return; a benefit. **2.** The return received on a business undertaking after all operating expenses have been met.

As we indicated above, this margin of profit is always calculated as a percentage of the selling price (generally unknown until calculated), while markup is applied as a percentage of an item's cost (always known once it is acquired).

Calculating Margin

Let's start with margin. As the proprietor of Above Average Automotive, you decide your business must have an overall (translated as *average*) gross profit margin of 50 percent *or more* on the parts you sell to be successful. To achieve that 50 percent margin of profit you must *divide* the acquisition cost of an item by 0.5 or *multiply* that cost by 2.0.

These are basically the same formulas we will use later on in Chapter 5 when we look at our KPIs.

Selling Price (or Parts Price) – Acquisition Cost (or Cost of Goods Sold) = Gross Profit on Parts

Gross Profit ÷ Selling Price = % Gross Profit

For argument's sake, we'll say the acquisition cost of the item in question is $45.81. We aren't sure what we are going to sell the item for as of yet, but we know we need a gross profit margin of 50 percent to *keep the boat afloat*—in order to achieve success. The invoice from the local parts house suggests the manufacturer's list for this part is $56.12. The difference between the cost of an item and its selling price is the gross profit associated with that sale. In this example, it is $10.31. If you *divide* the gross profit by the item's selling price, the result is your gross profit percentage, or margin.

In other words, if you *subtract* the $45.81 cost of the item from the MSRP of $56.12 you have a gross profit of $10.31. If you *divide* that $10.31 by the selling price—$56.12—the result is 18.3 percent, or a GPM of 18.3 percent.

We know that 18.3 percent is not enough regardless of what the manufacturer or the parts house suggests is appropriate. We know that we must increase that GPM to at least 40 to 50 percent to survive. To accomplish that, we can look at the margin calculator in Table 3-1 and determine what to do next.

TABLE 3-1 MARGIN CALCULATOR

To Achieve a Margin of:	Divide by:	or Multiply by:
10%	0.90	1.11
15%	0.85	1.18
20%	0.80	1.25
25%	0.75	1.33
30%	0.70	1.43
35%	0.65	1.54
40%	0.60	1.67
45%	0.55	1.82
50%	0.50	2.00
55%	0.45	2.22
60%	0.40	2.50
65%	0.35	2.86
70%	0.30	3.33
75%	0.25	4.00
80%	0.20	5.00
85%	0.15	6.67
90%	0.10	10.00
95%	0.05	20.00
100%	0.01	100.00

For a desired GPM of 40 percent, we move down the first column on the left to 40 percent and then across—we either divide by 0.60 or multiply by 1.67. Multiplying $45.81 by 1.67 will result in a suggested retail or list price of $76.50. Dividing by 0.60 will result in a retail or list price of $76.35. In this case, the difference between an 18.3 percent margin and a 40 percent margin is more than $20.00.

For a GPM of 50 percent, we would look at the chart and then either divide by 0.50 or multiply by 2.0. Dividing by 0.5 would result in a retail or list price of $91.62 and multiplying by 2.0 would result in a list or retail price of the same. In one case the difference in profit between the suggested retail price and our own is $30.69 and in the other, it is literally twice as much, or $45.81. The difference in bottom line profit is really significant when you realize that the additional $20.00 to $30.00 goes directly to your bottom line.

Calculating Markup

Since markup is calculated as a percentage of the cost of an item, we really have all the information we need. So let's say that a gasket set costs you $40.00. In order to *mark* the item *up* (markup) by 50 percent, you would take 50 percent of the cost of that item and then add it to that cost.

(% of Markup x Acquisition Cost) + Acquisition Cost = Retail or List Price

In this example, that would mean taking 50 percent of the $40.00, or $20.00 (your markup of 50 percent), adding them both together for a total selling price of $60.00.

(0.50 × $40.00) + $40.00 = $60.00

Using what we have just learned, the markup would be 50 percent (1/2 of $40.00, or $20.00), while the margin of profit would be one third of the selling price, or $20.00.

$20.00 ÷ $60.00 = 33.3% margin

To achieve a *margin* of 50 percent, we would need to mark up the part or accessory in question a total of 100 percent. This started to come together for me in Table 3-2. There you can see the relationship between markup and margin a little more clearly. The first column shows margin percentages increasing in 2.5 percent increments. The second column shows the percentage of markup required to achieve the same result. In all cases it is higher. The third column shows the calculation required to determine what the markup will be if you know what the desired margin is.

Table 3-2   MARGIN/MARKUP RELATIONSHIP CALCULATOR

A Gross Margin of:	Requires a Markup of:	Calculated:
10.0%	11.11%	10.0 ÷ (100 − 10.0)
12.5%	14.29%	12.5 ÷ (100 − 12.5)
15.0%	17.65%	15.0 ÷ (100 − 15.0)
17.5%	21.21%	17.5 ÷ (100 − 17.5)
20.0%	25.00%	20.0 ÷ (100 − 20.0)
22.5%	29.03%	22.5 ÷ (100 − 22.5)
25.0%	33.33%	25.0 ÷ (100 − 25.0)
27.5%	37.93%	27.5 ÷ (100 − 27.5)
30.0%	42.86%	30.0 ÷ (100 − 30.0)
32.5%	48.15%	32.5 ÷ (100 − 32.5)
35.0%	53.85%	35.0 ÷ (100 − 35.5)
37.5%	60.00%	37.5 ÷ (100 − 37.5)
40.0%	66.67%	40.0 ÷ (100 − 40.0)
42.5%	73.91%	42.5 ÷ (100 − 42.5)
45.0%	81.82%	45.0 ÷ (100 − 45.0)
47.5%	90.48%	47.5 ÷ (100 − 47.5)
50.0%	100.00%	50.0 ÷ (100 − 50.0)
52.5%	110.53%	52.5 ÷ (100 − 52.5)
55.0%	122.22%	55.0 ÷ (100 − 55.0)
57.5%	135.29%	57.5 ÷ (100 − 57.5)
60.0%	150.00%	60.0 ÷ (100 − 60.0)

In Table 3-3, I have tried to demonstrate how these relationships work with real numbers.

What we have done here is insert an imaginary part costing $40.00 into the chart in Table 3-2. To calculate the margin for that part you would take the desired percentage of margin, subtract it from 100 and then divide the part's cost by that number. So if our desired margin was 10 percent, we would subtract that 10 percent from 100 percent and then divide our $40.00 parts cost by 0.90.

$40.00 ÷ 0.9 = $44.44

And then so on down the chart. You can see the different margins and their impact on our $40.00 part as you move down the third column on the chart.

To achieve a markup equal to that margin we would use the same $40.00 part. However, our focus changes here. To calculate a corresponding markup equal to a certain margin, you begin by taking that margin and then dividing it by what is left when you subtract the margin from 100. The formula would look something like the one in the fifth column (identified as *Calculated*) of Table 3-3.

% of Margin ÷ (100 - % of Margin = % of Markup

or,

10.0 ÷ (100 - 10.0) = % of Margin

10.0 ÷ 90.0 = 11.11%

And the markup would be consistent with the number you will find at the top of the fourth column: a markup of 11.11 percent or $4.44. When you add that markup to the $40.00 parts cost, you have the same result as we achieved with a 10 percent gross profit margin—$44.44.

SUCCESS

In almost every survey of successful automotive service facilities we have ever seen, one of the primary signatures of a financially sound and viable business is a consistent margin of profit of 40 percent or better on all parts and accessories.

The only question is, is 40 percent enough? And the answer is, I'm not sure. There are too many variables to create a blanket rule. What do you do if a high percentage of your parts are purchased through the dealership network where the margins (synonymous with: *discounts*) are substantially lower than the discounts or margins available through the traditional automotive aftermarket?

If you buy parts at the dealership, you will generally receive a 20 percent discount (margin). That discount is based upon the manufacturer's suggested retail price. If you sell that part at the suggested retail price, your profit margin will be equal to the discount, or 20 percent (the amount of the discount), and your markup will be 25 percent. (You can see the percentages in Table 3-3, second and fourth columns, five lines from the top.)

TABLE 3-3   MARGIN & MARKUP RELATIONSHIP GUIDE WITH DOLLAR AMOUNTS AND CALCULATION FORMULA

Part Cost	Margin of:	List Price:	Required Markup:	Calculated:	List Price:
$40.00	10.0%	$44.44	11.11%	10.0 ÷ (100−10.0)	$44.44
$40.00	12.5%	$45.71	14.29%	12.5 ÷ (100−12.5)	$45.71
$40.00	15.0%	$47.06	17.65%	15.0 ÷ (100−15.0)	$47.06
$40.00	17.5%	$48.48	21.21%	17.5 ÷ (100−17.5)	$48.48
$40.00	20.0%	$50.00	25.00%	20.0 ÷ (100−20.0)	$50.00
$40.00	22.5%	$51.61	29.03%	22.5 ÷ (100−22.5)	$51.61
$40.00	25.0%	$53.33	33.33%	25.0 ÷ (100−25.0)	$53.33
$40.00	27.5%	$55.17	37.93%	27.5 ÷ (100−27.5)	$55.17
$40.00	30.0%	$57.14	42.86%	30.0 ÷ (100−30.0)	$57.14
$40.00	32.5%	$59.26	48.15%	32.5 ÷ (100−32.5)	$59.26
$40.00	35.0%	$61.54	53.85%	35.0 ÷ (100−35.0)	$61.54
$40.00	37.5%	$64.00	60.00%	37.5 ÷ (100−37.5)	$64.00
$40.00	40.0%	$67.67	66.67%	40.0 ÷ (100−40.0)	$67.67
$40.00	42.5%	$69.57	73.91%	42.5 ÷ (100−42.5)	$69.57
$40.00	45.0%	$72.73	81.82%	45.0 ÷ (100−42.5)	$72.73
$40.00	47.5%	$76.19	90.48%	47.5 ÷ (100−47.5)	$76.19
$40.00	50.0%	$80.00	100.00%	50.0 ÷ (100−50.0)	$80.00
$40.00	52.5%	$84.21	110.53%	52.5 ÷ (100−52.5)	$84.21
$40.00	55.0%	$88.89	122.22%	55.0 ÷ (100−55.0)	$88.89
$40.00	57.5%	$94.12	135.29%	57.5 ÷ (100−57.5)	$94.12
$40.00	60.0%	$100.00	150.00%	60.0 ÷ (100−60.0)	$100.00

Unless you are talking about high-dollar items like rebuilt or new engines and transmissions that are very expensive and consequently a price concern for the consumer, that 20 percent margin or 25 percent markup will not generate the dollars you will need to sustain your business. The reason to accept less in the case of those high-dollar items is that while the margins and markups are generally lower, the dollar amount you are working with is substantially higher, and the volume tends to make up for the shorter margins.

The question would have to be, why aren't you entitled to the same 40 percent margin of profit on dealership parts as you are on parts acquired through more traditional aftermarket channels? And the answer is, you are! In many instances, the dealership service departments are marking up the parts they are acquiring through their own parts departments, so why shouldn't you?

After all, it is you who will ultimately accept the liability for warranty and returns. Nevertheless, we must ask ourselves if parts are the only things the customers are buying when they come to us for automotive service, maintenance, and repair. If the answer is yes, and the cheapest price on parts is all the customers are looking for, they are in the wrong place! They can buy parts— a host of different parts—a lot cheaper at a host of other venues.

If, however, the customer is in the shop for all the right reasons, because they are interested in the quality of your service, your technical staff and parts, as well as all the additional value-added services you have to offer, a point or two of additional margin will not make much of difference one way or the other, as long as the vehicle is ready when promised and fixed right the first time—two things we will have lots more to say about in other guides in this series.

So the answer really depends as much upon the quality of your service and the quality of your customers as it does on the quality of your parts! The better the service, the better the customers... and the better the customers, the higher the margins because better customers are generally more service-sensitive than they are price-sensitive.

WHAT SHOULD I STOCK AND WHAT SHOULD I LET THEM STOCK?

When I started in this industry it was not uncommon for service dealers to have a significant amount of capital tied up in on-the-shelf inventory. In general, inventories were significantly higher than they are today—there were fewer applications and fewer part numbers. With the proliferation of part numbers we have seen in recent years much of that has changed. If you are not a *specialist*, it is virtually impossible to stock all the parts needed for normal service, maintenance, and repair on the mix of cars and trucks likely to appear at your gate in the course of a normal workweek. As a direct result of this phenomenon, the service industry has come to depend far more heavily on local jobbers and warehouses as a kind of *extended stockroom*.

The game has changed and so have the rules. Now, the most prudent thing you can do as an owner or manager with regard to stocking inventory is stock only those parts you are using frequently, the parts most likely to hang a job up or cause a serious delay. Non-critical parts that are stocked locally and remain readily available can and should be left on the shelf of the appropriate distribution resource and considered, as suggested earlier, *extended inventory*.

Just remember that a distribution partner that provides such a service is not doing so without incurring the same costs you are saving by not having those same parts on the shelf in your stockroom—costs that must be reflected in the prices you will be asked to pay when those parts arrive.

As far as the capital you do have invested in on-the-shelf inventory is concerned, just remember that money invested in inventory can be a better investment than money invested any other way-as long as that inventory turns. How can you tell if your inventory is turning? You look at what you have on hand, what you have purchased, and what you have sold, and those relationships will tell you how much of your inventory is turning and how much is standing still. In Chapter 2, we looked at this concept and defined it.

Inventory turns or *turnover* refers to how fast those parts and accessories you have on the shelf in the back room are being removed for installation, then repurchased and returned to the shelf to be sold again. It is calculated by dividing your cost of goods sold (per year) by your average inventory (per month). The equation looks like this:

Cost of Goods Sold Per Year ÷ Average Inventory = Inventory Turnover

What Do You Do When Inventory Doesn't Turn?

Certainly that question is infinitely easier to answer today, with the computerized shop management systems that are out there, than it has ever been before. Most of these systems have fairly sophisticated inventory modules with some fairly impressive report generating capabilities. Those reports should be able to tell you exactly what is moving and what is not. If you don't have a computerized shop management system, or you do but it doesn't have an inventory management module that can do this for you, there are and have always been a number of other ways to track inventory sales.

One of the ways to determine how fast inventory is moving, or even whether it is moving at all, is to place a small adhesive dot on every box, can, and carton in your stockroom, with the date it was entered into inventory on the dot. To track movement, all you have to do is reconcile the date the part sold with the date it was purchased. If you wanted a little more sophistication, you could use dots that were marked with twelve different colors—one for each month of the year. Or you could attach an inventory card to each item.

Whatever you do will require an investment of one kind or another. It will require measurement and monitoring, and that will take either your own personal time and manual effort or an investment in the kind of computerized technology mentioned above. Either way, there will be a cost.

The return on that investment should be a smaller inventory that turns more frequently, increased profits, and fewer returns.

JOBBERS, WAREHOUSES, SHORT-LINERS, AND SPECIALTY WAREHOUSES

There has been a fairly dramatic shift in the distribution industry that started more than thirty years ago. What was once known as traditional three-step distribution, a system that was defined by the movement of manufactured parts through the distribution system from the manufacturer, to the warehouse and then on to the jobber store, is being displaced by a number of different yet similar responses to changes in the marketplace.

Some jobbers grew large enough to buy direct, reducing their cost and increasing their profits. Other niche players entered the marketplace as specialists, competing in only one area of automotive parts and products such as steering, suspension and undercar, or European or Asian imported car parts. Some warehouses decided they would only handle certain select, fast moving parts, leaving the balance of harder to move parts and accessories to the traditional full-service warehouses, while others opted to bypass the jobber altogether, moving parts directly from the manufacturer to the service dealer.

Specialists, short-liners, and two-steppers share a great deal in common. Almost all offer limited inventory and limited stock. In other words, they don't always have what you need when you need it. Although their inventory is often lacking, they compensate with attractive and aggressive prices, which are often passed on to you as their service dealer customer. But there is a cost—a downside—and that downside is service.

If the consumer's primary concerns are *ready when promised* and *fixed right the first time*, where do limited lines and abbreviated inventory figure into the equation? You can't get the vehicle finished on time if you can't get the parts you need when you need them. And you certainly can't offer the vehicle owner the kind of confidence and security they want and need without access to branded merchandise and the best manufacturers. Consequently, there is still an argument to be made in our industry for the full-service warehouse. Having the right part in the right place at the right time has value, or at least it should! Being able to have the vehicle ready when promised with the kind of parts you can feel confidence in has value as well.

The cost of parts delivery has increased to over $7.00 every time a jobber store or warehouse driver puts a key in the ignition switch of a delivery vehicle. That is $7.00 regardless of what you have ordered. Like you, they must recover that cost somehow. Try to remember that when you pick a distribution partner.

As in all things, the real issue here is balance—balance, and knowing what both you and your customers want and need. Everything tells us that when your customers choose a facility for automotive service and repair, it's a matter of ready when promised, fixed right the first time, and ultimately value. That translates to service and availability. There really isn't much difference when the choice is ours and we are picking a distribution partner. What you are really interested in is a combination of service, availability, and price. If you can find that, embrace it. It is as valuable as it is rare.

SUMMARY

For some reason, margin and markup are two of the most difficult business concepts for most shop owners to understand. Margin is calculated as a percentage of the selling price of an item while markup is calculated on the purchase price or acquisition cost.

Because margin is calculated on the selling price of an item, and because the selling price of an item is always higher (or at least it should be!) than the purchase or acquisition cost, and because markup is always calculated on the purchase or acquisition cost, a 40 percent margin will produce a higher gross profit than a 40 percent markup will.

Understanding the difference between margin and markup will help insure that you are getting the highest return on the sale of both labor and parts.

Understanding the different ways parts move through the distribution channel will help you pick the right distribution partner, just as understanding what you should stock in inventory and what you should buy out will help you maximize your profits-not only when it comes to the parts you buy, but also in regard to managing service bay productivity, because you can't complete a job until you have the appropriate parts.

What will you do to manage your inventory better? What will you do to insure adequate margins and markups? Write down some of your ideas in the chart on the next page.

Thoughts

Actions

Results

CHAPTER 4

Using What You Know

INTRODUCTION

This chapter introduces you to:

- Using your cost of doing business to establish your labor rate

- Setting prices for services based on your needs, not the needs of your competition

It will help you use what you know to price your products and services more intelligently—without having to rely on Oreo pricing, instinct, intuition, or just plain guessing. It will explore the danger and fallacies involved in:

- Pricing below other service providers in your area in order to create a false sense of value

- Using the labor guide exclusively to set labor times

It will help develop alternative methods to establish pricing policies on the basis of:

- Cost of doing business

- Personal profit demand

- Reverse budgeting

- Recovering expense

DANGER!

Beware Oreo Pricing—the tendency of automotive shop owners and managers to price their products and services between what they perceive to be the highest and lowest priced service providers in their market area. There is an inherent danger here. It's too easy to hide in the soft creamy center of the cookie! Pricing your products and services on the basis of what others have done or continue to do can be hazardous to the long-term financial health and well being of your business. Aside from that, it presupposes that your competitors have done the math and their homework and have set prices where they are legitimately supposed to be, based upon their Cost of Doing Business (CODB). It presupposes that they have an idea of what their products and services cost them. Perhaps more importantly, it assumes that they know what they need to charge in order to insure an adequate return on their investment. It presupposes that they understand what is necessary based upon sound business practices and that they have the courage to price accordingly. It also assumes that your competitors have not established their pricing strategies the same way you have—based upon what they think you are doing in your shop!

Does anyone really know who the true market leader in your neighborhood is anyway?

Assumption I: A Couple of Dollars Less Than...

Perhaps the most dangerous and destructive assumption of them all is what I call the *A Couple of Dollars Less Than* syndrome, and it manifests itself in the way most shop owners consciously set their prices to insure they are pricing just below those prices posted at the factory dealerships in their areas. The rationale for pricing beneath the factory service outlets is clear even if the explanation isn't-in the mind of the independent it creates an artificial value differential. Too many shop owners act as if the only way they can compete, the only way they can create enough value to make them an attractive alternative to the dealership service department, is by creating an image of equal quality with dramatically lower prices.

This is particularly dangerous because it assumes that the dealership service department is somehow better, more able, more talented, more sophisticated, and more apt to meet or exceed the consumer's wants, needs, and expectations. This may or may not be true. It depends as much upon the individual dealership service department and the way it is run as it does on the qualities of the specific independent garage in question.

The A Couple Dollars Less syndrome takes away the possibility of leadership before the competition has even had the chance to begin. It subliminally admits inferiority. And it suggests that the only reason for patronizing an independent is somehow related to price and price alone.

Most frustrating of all is the fact that most dealership service departments cannot raise their prices until there is an increase in market area pricing that can be confirmed by an actual audit. By pricing below the dealership service departments, most independents are actually shooting themselves in the foot. By not allowing the dealerships' prices to go up, they prevent themselves from enjoying the prices and the profits they have worked so hard and risked so much to achieve.

Our shop posts a higher labor rate than virtually all of the dealerships in our community. We work at keeping our prices at the top of the market and have for almost twenty years. At one point we charged as much as twenty dollars an hour more than the seven dealers who have subsequently gone out of business in our community. If we had accepted the traditional paradigm and priced our labor and parts below those dealerships, we might have found ourselves sitting right beside them in bankruptcy court!

Remember, the dealership you may be monitoring to insure that your prices are competitive may not even know what it costs *them* to remain in business, let alone what *your* expenses, personal needs, and profit demand might be.

Assumption 2: Price is the Motivating Factor in a Decision to Choose One Automotive Service Facility over Another

Ready when promised and *fixed right the first time* are the two most commonly articulated reasons vehicle owners offer when asked what they are looking for in automotive service. They are consistently the number one and number two answers in every survey of customer wants, needs, and expectations we've seen over the past two decades.

Where does that leave price? No higher than third on anyone's list, and not even there when you throw other customer wants, needs, and expectations into the mix—they also want convenience, quality, responsiveness, and accountability!

What is the key factor if it isn't price? It is a combination of many different dynamics, but it has more to do with personal service, intimate relationships, convenience, quality, and meticulous attention to detail than anything else. People come to independent shops for two of the most powerful psychological reasons in the world: recognition and appreciation. Don't ever forget that-it's why most people choose one facility over another when patronizing any kind of business for any kind of product or service.

Assumption 3: The Labor Guide

Most shop owners act as if the labor guides they use to estimate the work they do were delivered along with the two tablets of the law Moses brought down with him from Sinai!

I believe the industry is too dependent upon the guides for estimating and final billing. We should remember that the labor guide is nothing more or less than what it says it is: a *guide*. It is one tool, to be used in conjunction with a number of other tools in your estimating and invoicing toolbox. It is not the only tool in the toolbox, and we must remember that.

Most shop owners are aware that the guides are not always accurate, and yet most shop owners never challenge that accuracy or develop alternative strategies to base their labor estimates on. Certainly there is a place for comparison-shopping in the estimating and invoicing toolbox. Almost every other industry resorts to some kind of intelligence gathering when it comes to competitive pricing—why not us? Why not just pick up the phone and call another shop, a shop you believe delivers a comparable level of quality and service, to find out what someone else, someone you know, respect, and trust might charge? Or, perhaps just as important to find out what consumers in your market area have accepted as reasonable and competitive with regard to any specific automotive service, maintenance operation, or repair.

There is a place for experience in this process—a place that would allow a trusted technician to estimate how long it should take to do the job, as well as what it would take to do the job in a way that would make everyone in your company proud. There is a place for added value in this equation—a place for care and for quality, for dedication and for concern. There is a place for a warranty and, perhaps more importantly, the policies and procedures you have established to honor that warranty. And finally, there is or at least should be a place for a discussion of what a job is worth—worth to you and worth to the consumer.

I no longer feel compelled to allow the labor guide to fix my labor prices, any more than I feel compelled to allow the local parts house, warehouse, or manufacturer to set my retail prices on parts or accessories. They have not yet proven that they understand my business, my market, or me well enough to do that effectively and successfully. They don't seem to understand or appreciate the cost of a highly skilled and qualified technician any more than they understand the investment required to secure high-tech equipment, automotive service education and training, or many of the other costs involved in operating a successful automotive service business today.

We must also remember that any tool can be misused, used improperly, or not used effectively for any one of a number of reasons. I have seen this all too often, especially with regard to the *adds* that fail to find their way onto a customer's estimate. These adds accompany virtually every removal and replacement operation in the labor guide. They are the 0.2 hours if the vehicle is equipped with electronic engine controls or the 0.5 hours for power steering, the 0.2 hours for each additional core plug and the 0.5 hours to replace cam seals with a timing belt. The electronic guides at least force you to confront additional charges. They force you to either consider them or ignore them, but they can't force you to include them on the estimate.

There are far too many shop owners who stop the estimating process with a quick look at the labor time for removal and never look at or include the adds.

The Bottom Line

Oreo pricing is dangerous and deceptive. It gives too many garage owners a false sense of security and allows too many others the luxury of just *winging it* without analyzing the costs or profit demands required to establish a price that works for you as well as for the customer.

I wouldn't want the pilot on a transcontinental flight to just wing it on his or her way to New York, and I wouldn't want my dentist to just wing it the next time she lets her fingers do the walking inside my mouth. I want someone with experience, education, training, and some kind of plan working for me!

THE WAY TO DO IT

You must believe that you are entitled to a profit on the products and services you offer your clients—a profit that reflects the risk and responsibility that accompanies entrepreneurship. There are far too many shop owners who act as if what they are already making is somehow too much.

No one is ever going to confront you demanding that you raise your prices in order to insure that you remain in business. It's just not going to happen! No one has ever stormed into my office insisting my prices were too low—even when they were!

You start with a number—a dollar amount sufficient to satisfy your need for financial security and grow your business at the same time. Do the analysis, then do the math.

If the prices you are charging aren't justified by the quality of service you are providing, it's time to re-evaluate more than just your prices.

We will start with pricing our labor more intelligently because that is where we can realize the most dramatic and immediate impact and improvement. It is also something we should be able to measure and manage without great difficulty.

Rather than allowing ourselves to become dependent upon what our neighbors are or are not doing, we are going to set our prices based upon our actual cost of doing business per hour factored for service bay productivity. We will spend a lot more time explaining this method of determining your hourly labor rate in the next section, but for now we will start by totaling all of the expenses involved in the operation of your business, with the exception of cost of goods sold. Cost of goods sold is exactly that—the cost of the parts and accessories you purchase and then resell in the course of performing automotive service. Once we have that total, we can either distribute it over every hour you are open for business or over every hour you bill. We will be factoring that number by the average service bay productivity of your shop's technical service staff to help establish a labor rate, so we'll stick to billed hours. This is critically important because, as we have seen, you don't always invoice every hour you have available to invoice. Consequently, productivity has a profound impact on profitability and should have an equally profound impact on what you must charge in order to insure an adequate return on investment.

If this total cost of doing business is not recovered, you will not generate a profit. If revenues don't exceed your costs, you will experience a loss. To avoid that, you can add a percentage of profit to every hour you invoice. You have to know the costs associated with remaining open, and those costs must be factored by your average service bay productivity. Then, and only then can you determine if your prices are reasonable and competitive for your market area.

Once you have established an hourly labor rate factored for service bay productivity that reflects your cost of doing business *without* cost of goods sold, you can turn your attention to the issue of profit on parts. As you will see, the margin of profit on parts cannot be overstated. Too often, that margin of profit on parts may be all that is keeping the shop open and the owner in business.

This is not something new. Garage owners have consistently subsidized automotive service through a distinct lack of knowledge and understanding. They have absorbed the costs involved in performing repairs and services with artificially low labor charges. All too often, profit in the automotive service environment has been solely the result of the profit margin on parts.

By now, we should have finally realized that we cannot subsidize automotive service with artificially low prices on either labor or parts. The continually escalating cost of education, training, equipment, environmental responsibility, and governmental regulation has dictated that a reasonable profit on the parts we sell—an average margin of 45 percent or higher, in conjunction with an adequate margin of profit on labor, is critical to survival. That will make gross profit on both parts and labor one of the primary KPIs for monitoring success. In our business that means adequate margins on both.

MARKET LEADERS

Intelligent pricing is a matter of good management, clear vision, and thoughtful, effective leadership. It means performing the kind of work that insures good margins on both parts and labor. It means either acquiring the training, education, and equipment necessary to perform effectively, or abandoning that area of performance to someone who can do it better. It means recovering or reducing expenses while increasing sales and keeping margins high.

The market leaders in your industry choose the areas in which they compete carefully. Their prices reflect the investment they have made in tools, training, technology, equipment, and personnel. They understand the changes taking place in our industry, especially the acute shortage of qualified and competent technicians and the investment that must be made to leverage the ability of those technicians who remain. They anticipate and accept change as a natural part of their future and recognize that with change comes opportunity. Market leaders know exactly what their products and services cost and how much margin it will take to earn a profit. They have made the commitment to do what they do well. But in the process, they have managed to get paid well for doing it.

Market leaders understand the costs involved in doing business as well as the investment required to remain profitable. They understand that this investment can only come from one place, from one source, and that source is the vehicle owner. Market leaders not only understand the need to charge for diagnostic time, they also understand the need to charge more for diagnostic time because of the revenue that is lost when there are no parts sales to accompany the labor charges because all you are doing is inspection and testing.

A Reasonable and Competitive Price

Perception is reality. In our world, the only thing that matters is the customer's perception of us and our business. In the end, we are all governed by the consumer's perceptions of quality of service and of a reasonable and competitive price. My research has revealed that a reasonable and competitive price is a price that falls within an eight to twelve percent range of what the vehicle owner will accept as *normal and customary*.

Unfortunately, that can be deceiving in our industry. After all, what is a tune-up, and how complete is a *complete* cooling system service?

Is the vehicle owner capable of comparing apples and oranges? Who will explain the difference? As the professional in this relationship, it is your responsibility to aggressively research and comparison-shop the other service providers in your market area, just as it is your responsibility to educate your customers about the quality and value-added service you are able and willing to provide.

Market leaders consistently achieve the highest returns on their investments because they make a habit of good business practices. They both know and understand who their customers are. They know and understand what those customers want, need, and expect from the service industry. They realize that there is a difference between the tangible commodities they sell and the intangible product the vehicle owner wants to purchase. Market leaders push the pricing envelope every day, but they also search for new and innovative ways to add value. They are aware of the competition, but do not allow other service providers to dictate how they will do business.

What does it take to be a market leader? I'm not sure I have all the answers yet, but I have a much better understanding of what it takes today than I did some twenty years ago when I started this odyssey. We have traveled far and learned a great deal. We have become far more profitable than we once were, and yet I still sense there is infinitely more to learn.

It is critical that we continue to develop a better understanding of the dynamics that govern the marketplace as well as a deeper understanding and appreciation of what our customers want, need, and expect from us. It is just as critical for us to remain aware of our marketplace and the other service providers who share that marketplace with us. In the end, no one can hide in the sweet, creamy center of the cookie. It isn't safe.

To be truly successful you must continually invest in your quality of service, and that takes capital. If a customer, any customer, is willing to pay a higher price for the same service someplace else, we must understand what we have to do to make that customer comfortable spending that same amount with us.

Market leaders know how to do that. They are never comfortable with the status quo, never satisfied or complacent. You will always find them out in front, never lost in the soft, indistinguishable, creamy center of the cookie.

ESTABLISHING A LABOR RATE WITHOUT OREO PRICING

In the previous chapter we talked at great length about how *not* to establish a labor rate and then touched on one alternative. We're going to take an in-depth look at that alternative now, so that hopefully you won't be tempted to take the easy way out and *Oreo price* your parts and labor in the future. Real world conditions may make it difficult to put into practice the rules and the lessons you have mastered in theory, and there are forces in the market place that can make it difficult if not impossible to set your labor rate purely on the basis of a cost of doing business calculation. But that does not diminish the importance of knowing how and when to make this calculation.

To provide you with the tools you need to be successful, I will be explaining a simple and effective formula for calculating an hourly labor rate based upon your own cost of doing business that is particularly effective because it is factored for *service bay productivity*. The purpose for the exercise is to help you establish a retail hourly labor rate without having to resort to Oreo pricing.

In virtually all the seminars I have presented, the one thing that has surfaced most consistently is an almost unreasonable unwillingness on the part of shop owners and managers to become intimately involved with the numbers that *are* their respective businesses. There is an honest and genuine reluctance to confront pricing issues and what must be charged for both parts and labor services in order to get a legitimate return on investment.

As a service dealer with more than thirty years of experience, that is certainly something I can relate to. I resisted dealing with those very same numbers as long as I was able to. The only problem with this approach to business management is that it left the management part of the equation out! You can't really manage anything other than workload if you don't at least start to pay attention to the numbers. And you will never be successful at working smart and not just working hard until you take the time and make the effort to embrace and understand the numbers.

COST OF DOING BUSINESS

Cost of Doing Business (CODB) calculations should be made at the earliest opportunity. Otherwise, you will find yourself doomed to price your goods and services along with all the other garage owners who still insist on charging a couple of dollars an hour less than the dealership and in line with everyone else on the street-the majority of whom still have no idea how or why their prices are where they are.

There is only one problem with the *couple of dollars less than the dealership* pricing philosophy-it isn't necessarily all that competitive. It presupposes that you know exactly what the dealerships charge and that requires a fairly sophisticated level of comparison-shopping and intelligence gathering. And it also requires an in-depth awareness of exactly where your neighbors are pricing their goods and services-something equally difficult. Oreo pricing assumes that your competitors are *not* doing the same thing you are and that's not often the case.

So, What Is It?

The CODB is exactly that—all of the costs related to your being able to provide whatever service or services you offer. It is *all* of your costs: tools, rent, insurances, licenses, taxes, permits, fees, hazardous waste disposal, management and containment, management, training, advertising, lights, heat, etc, with the exception of cost of goods sold (more about that in a minute), factored by units of time and then by average service bay productivity.

Despite the fact this is a fairly long definition, it does tell you everything you need to know about what you are about to do. What it doesn't tell you is how you're going to do it.

The first thing you do is get in touch with your numbers and that means accumulating your financial statements over the longest time period available in which business has been fairly consistent. This is a good place to get in touch with your accountant. Your accountant should be more than willing to help you understand your financials better. However, there is always the possibility you might experience some resistance when it becomes apparent that you want or need better financial counseling and a lot more information than you have been satisfied with in the past. The more you realize you need to know, the more you grow—the better your questions, the more difficult your bookkeeper or accountant's job is likely to become, at least for a little while.

Information is power. So the deeper the data, the more accurate your calculations will be. Three years of history would be ideal. That will give you enough background to factor out the unexplained peaks and valleys, surges in volume, anomalous expenses, or glitches in the economy.

If you don't have three years of history available, don't wait until you do. It is critical to achieve an accurate idea of just what it costs you to open the doors in the morning! Take the numbers you have available and fold them into the following templates. Any sense of your actual CODB, even one that is not as accurate as it could be, is still better than Oreo pricing.

Table 4-1 is the sample annual income statement (P&L statement) from Above Average Automotive that we are going to be using for our examples.

TABLE 4-1   SAMPLE ANNUAL P&L STATEMENT FROM ABOVE AVERAGE AUTOMOTIVE

Sales	12 Months to Date	Total Sales (%)
Parts	$155,505	37.11
Outside Labor	$15,761	3.76
Labor	$249,451	59.53
Discounts and Refunds	($2,109)	-0.50
Miscellaneous Income	$439	0.10
Total Sales	$419,047	100.00
COST OF GOODS SOLD		
Direct Cost		
Direct Labor	$76,964	18.37
Purchases: Parts	$118,145	23.74
Outside Services	$9,552	2.28
Less Ending Inventory	($12,949)	−3.09
Less Ending WIP	($15,259)	−3.64
Total Cost of Sales	$99,489	23.74
Indirect Cost		
Depreciation	$13,656	3.26
Rent	$45,600	10.88
Repairs and Maintenance	$4,726	1.13
Operating Supplies	$759	0.18
Shop Expenses	$21,217	5.06
Total Indirect Cost	$162,922	38.88
Total Cost of Goods Sold	$262,411	62.62
Gross Profit	$156,636	37.38
OPERATING EXPENSES		
General and Administrative		
Accounting	$2,427	0.58
Advertising	$1,708	0.41
Alarm Security	$108	0.03
Auto Expense	$945	0.23
Auto Repairs	$336	0.08

TABLE 4-1   SAMPLE ANNUAL P&L STATEMENT FROM ABOVE AVERAGE AUTOMOTIVE (CONTINUED)

General and Administrative	12 Months to Date	Total Sales (%)
Bad Debts	$2,92	0.70
Bank Charges	$246	0.06
Consulting Fees	$995	0.24
Credit Card Discounts	$1,904	0.45
Dues and Subscriptions	$1,270	0.30
Equipment Lease and Rental	$10,068	2.40
Interest Expense	$2,434	0.58
Insurance: General	$5,271	1.26
Insurance: Group	$9,883	2.36
Insurance: Workers Compensation	$6,758	1.61
Laundry and Uniforms	$2,174	0.52
Meals	$86	0.02
Office Supplies	$6,577	1.57
Office Salaries	$8,387	2.00
Officer's Salaries	$60,231	14.37
Payroll Tax Expense	$11,383	2.72
Penalties	$121	0.03
Refunds	$604	0.14
Taxes and Licenses	$5,230	1.25
Telephone	$4,979	1.19
Utilities	$5,438	1.30
Total General and Operating	**$152,484**	**36.39**
Net Income Before Taxes	**$4,152**	**0.99**
Provision for Taxes		
Provision for Franchise Tax	$800	0.19
Provision for Federal Tax	$509	0.12
Total Tax Provision	**$1,309**	**0.31**
Net Income	**$2,843**	**0.68**

Table 4-2 is the sample spreadsheet we pulled out of the statement above showing all the expenses that might be included in a CODB calculation with the exception of cost of goods sold (the expense involved in purchasing the parts and accessories which we acquire only to resell).

TABLE 4-2 CODB—LIST OF EXPENSES FOR THE PREVIOUS 12 MONTHS

	12 Months to Date
Direct Cost	
Direct Labor	$76,964
Outside Services	$9,552
Total Cost of Sales	$86,516
Indirect Cost	
Rent	$45,600
Operating Supplies	$759
Shop Expenses	$21,217
Total Indirect Cost	$67,576
Total Cost of Goods Sold	$154,092
Operating Expenses **General and Administrative**	
Accounting	$2,427
Advertising	$1,708
Alarm Security	$108
Auto Expense	$945
Auto Repairs	$336
Bad Debts	$2,921
Bank Charges	$246
Consulting Fees	$995
Credit Card Discounts	$1,904
Dues and Subscriptions	$1,270
Equipment Lease and Rental	$10,068
Interest Expense	$2,434
Insurance: General	$5,271
Insurance: Group	$9,883
Insurance: Workers Compensation	$6,758
Laundry and Uniforms	$2,174
Meals	$86

TABLE 4-2 CODB—LIST OF EXPENSES FOR THE PREVIOUS 12 MONTHS (CONTINUED)

Operating Expenses General and Administrative	12 Months to Date
Office Supplies	$6,577
Office Salaries	$8,387
Officer's Salaries	$60,231
Payroll Tax Expense	$11,383
Penalties	$121
Refunds	$604
Taxes and Licenses	$5,230
Telephone	$4,979
Utilities	$5,438
Total General and Operating	$152,484
CODB for 12 months	$306,576

As you can see, we have omitted any costs related to parts or accessories. Since we are going to use this formula to help us determine our hourly labor rate and we really don't know whether or not we will sell any parts or accessories when we open the doors, we're going to leave parts and accessories out of the equation for now.

Are there any other costs that should be included in the above list, something that may not appear on the spreadsheet, something specific to your business? I don't know. I think I have included everything pertinent to a business like ours—every cost or expense I think would have an impact on this calculation. However, only you and your accountant will know if you have included everything that you need to in order to insure an accurate result for you.

There is one caveat here. If you find yourself including expenses that are not legitimately a part of your cost of doing business, personal expenses that somehow get paid by the business, factor them out of the equation for the sake of accuracy. They are not a genuine cost of doing business and they will corrupt the results.

Yearly CODB Expense

When we plug in all the numbers, we have a business that incurred $306,576 in expenses for the year, not including the purchases of parts. Once we have established the total expense, we can start breaking that total into manageable units.

We will allow Above Average Automotive, the imaginary business we have created serve as our example. As we move ahead with these calculations, we will have to establish a set of data points, givens—things like hours of operation, number of days the business is open per year, the number of technicians, and how efficient or productive they are. We can't go any further until we have established just what these givens are. So let's do that now. Above Average Automotive is open five days a week, fifty-two weeks a year, and is staffed by 3.8 technicians. We will be varying the service bay productivity for this business in several different examples to illustrate just how dramatically productivity can impact your bottom line. But, for now, average service bay productivity for Above Average Automotive is just barely above average at 63 percent.

How can you have 3.8 people working for you? Actually, it's easy. You might have an entry-level technician who is only capable of working a fraction of the hours expected of your more seasoned, veteran technicians. Or you could have a technician who is involved in a work/study program and is only in the shop six hours per day. In other words, there are a lot of different ways you can account for a partial technician. All you have to do is know what those ways are and whether or not they apply to you.

Daily CODB Expense

Fifty-two weeks a year times five days per week results in 260 working days for the year. If you look at our example in Table 4-2, you will see that our total cost of doing business for the 12-month period under review is $306,576.

As you can see in Table 4-3 below, if you divide that $306,576 figure by the 260 possible working days, our average CODB per day is $1,179.

TABLE 4-3 DAILY CODB CALCULATIONS

Days Open per Week	5
Weeks Open per Year	52
Working Days per Year	260
CODB for 12 months	$306,576
CODB per Day	$1,179

Remember that it isn't wise to sacrifice accuracy for the sake of expediency. It is critical that you plug in the right numbers for your business when you do this exercise. That means that if your shop is open a half-day on Saturdays as well as the five days per week we have counted, you need to add the appropriate number of additional days to your calculation. That would mean an additional 52 half-days or 26 additional working days. What about holidays? How many are you open and how many are you closed?

The important thing is to ensure that these numbers—the numbers you use—accurately reflect the number of days you are open, the number of hours you are open, and the actual number of technicians on-staff—not just the examples we've used here.

We now understand why the proprietor of Above Average Automotive is so anxious when he gets to the shop every morning. He knows that by the end of the day, he must generate $1,163.69 just to satisfy the daily expense involved in running the shop, or suffer a loss for that day!

Factoring for Technician Productivity

The easiest thing to do now would be to divide the total number of billable hours into that $1,179 figure in order to identify the cost of doing business per technician hour. However, there are some problems with that. The first problem is the formula—the number of billable hours equals the number of hours available to be billed per technician times the total number of technicians, because it's not accurate for most shops.

Number of Technicians x Hours per Day x Days per Period = Total Billable Hours

That means if you have two technicians working eight hours per day at 100 percent productivity (billing every hour you have available to bill) you have sixteen hours to invoice. Above Average Automotive employs 3.8 technicians. So you would insert that 3.8 into the above equation when multiplying the number of days times the number of hours per day.

3.8 Technicians x 8 Hours per Day x 260 Days per Year = 7,904 Technician Hours per Year

But are you billing one hour for every hour the technician is on the shop floor? In other words, are all or any of your technicians 100 percent productive? Are they invoicing every hour there is to invoice? And what impact will that have on your calculations if they are not?

Productivity has a profound impact on the cost of doing business. It has the potential to drive your cost of doing business up or down faster than any other single factor! You can certainly influence your overall CODB by monitoring or managing any one of the expenses that appears there. Serious improvement, however, will only materialize once service bay productivity percentages begin to improve.

The reason is simple. You only have so many hours to invoice. The more efficient your technicians, the more productive the shop, the lower the costs associated with actually doing that business will be when factored for increased productivity.

In our 40 hour per week model, the total number of billable hours per technician is 2,080. That is 40 hours per week times 52 weeks per year, as we can see in Table 4-4.

TABLE 4-4   WORKING HOURS PER YEAR FORMULA

Days Open per Week	5	Working Days per Week
Hours per Day	8	Hours per Day
Hours per Week	40	Hours x Days
Weeks Open per Year	52	Working Weeks
Working Hours per Year	2080	Hours x Weeks

A technician would be considered 100 percent productive if she could invoice every one of those 2,080 hours per year. At 100 percent productivity, 3.8 technicians could bill 7,904 tech hours a year. Under those circumstances, our spreadsheet for cost of doing business per hour would begin to look like the one in Table 4-5.

TABLE 4-5 CODB PER TECHNICIAN HOUR FACTORED BY 100 PERCENT SERVICE BAY PRODUCTIVITY

CODB 12 months	$306,576	
Days Open per Week	5	Working Days per Week
Hours per Day	8	Hours per Day
Hours per Week	40	Hours x Days
Weeks Open per Year	52	Working Weeks
Working Hours per Year	2080	Hours x Weeks
Number of Technicians	3.8	Technicians
Total Possible Tech Hours per Year	7904	Possible Hours per Day x Days per Year x Techs
Divided by:		
Total Possible Tech Hours	7904	
CODB per Hour @ 100 percent SBT	$38.79	

What you see above is the overall cost of doing business per technician hour of Above Average Automotive, based upon every technician in the organization working at 100 percent productivity. That means each technician is billing every hour that is available to bill, and that you as a service salesman, manager, or owner are selling every hour available for sale.

Our CODB per hour factored by an average service bay productivity of 100 percent (everyone billing every hour possible for them to bill) is $38.79. Anything less than 100 percent productivity will drive that $38.79 number higher. Realizing that, task management, technician efficiency, and measuring service bay productivity take on a whole new meaning and sense of urgency!

How many shops are there where every technician is 100 percent productive hour after hour, day after day, week after week? In most shops, the productivity curve is not distributed evenly. A brake and front-end specialist can consistently invoice 110 percent or 120 percent of the hours available. But what about the drivability expert struggling to bill 80 percent of his or her time because the owner is unable or unwilling to charge the customer for the time it actually takes to determine what is wrong? What does that do to average service bay productivity?

After hundreds of automotive shop management seminars, I can tell you that most shops are only billing slightly more than one hour out of every two available! If my experience is a true measure of the reality most shops face, average service bay productivity is hovering just above 50 percent.

To demonstrate just how critical high service bay productivity numbers are to CODB calculations, I'd like to show you what happens when the only thing we change is productivity.

Remember that service bay productivity is a calculation of the hours billed in relation to the hours available to be billed expressed as a percentage or ratio. At 87.5 percent productivity—seven hours billed out of every eight hours available to be billed—the spreadsheet starts to look like the example in Table 4-6.

TABLE 4-6 CODB PER TECHNICIAN HOUR FACTORED BY 87.5 PERCENT SERVICE BAY PRODUCTIVITY

CODB 12 months	$306,576	
Days Open per Week	5	Working Days per Week
Hours per Day	8	Hours per Day
Hours per Week	40	Hours x Days
Weeks Open per Year	52	Working Weeks
Working Hours per Year	2080	Hours x Weeks
Number of Technicians	3.8	Technicians
Total Possible Tech Hours per Year	7904	Possible Hours per Day x Days per Year x Techs
Service Bay Productivity (%)	87.5	SBT (%)
Divided by:		
Actual Tech Hrs per Year x SBT (%)	6916	Actual Tech Hours per Year x SBT (%)
CODB per Hour x SBT (%)	$44.33	CODB per Hour Factored x SBT (%)

By failing to bill one hour per day per technician we have increased our CODB per hour from $38.79 to $44.33. That is an increase of $5.54 per technician per hour, or $21.05 per hour for all 3.8 technicians!

This is why service bay productivity is the one KPI that almost every industry expert agrees is the most critical to the success of your business. Unfortunately, there are relatively few shops that achieve 87 percent productivity! Most shops operate below the 75 percent range and, as you can see in Table 4-7, those numbers would escalate your cost of doing business per hour to $51.72 per technician hour.

TABLE 4-7 CODB PER TECHNICIAN HOUR FACTORED BY 75 PERCENT SERVICE BAY PRODUCTIVITY

CODB 12 months	$306,576	
Days Open per Week	5	Working Days per Week
Hours per Day	8	Hours per Day
Hours per Week	40	Hours x Days
Weeks Open per Year	52	Working Weeks
Working Hours per Year	2080	Hours x Weeks
Number of Technicians	3.8	Technicians
Total Possible Tech Hours per Year	7904	Possible Hours per Day x Days per Year x Techs
Service Bay Productivity (%)	75.0	SBT (%)
Divided by:		
Actual Tech Hours per Year x SBT (%)	5928	Actual Tech Hours per Year x SBT (%)
CODB per Hour x SBT (%)	$51.72	CODB per Hour Factored x SBT (%)

For argument's sake, we'll say that everyone on our model shop's management team has read every book in this series and as a result the shop is a paradigm of efficiency and productivity consistently invoicing 85 percent of the shop's available hours. The numbers in Table 4-8 tell the story.

TABLE 4-8 CODB PER TECHNICIAN HOUR FACTORED BY 85 PERCENT SERVICE BAY PRODUCTIVITY

CODB for 12 months	$306,576	
Days Open per Week	5	Working Days per Week
Hours per Day	8	Hours per Day
Hours per Week	40	Hours x Days
Weeks Open per Year	52	Working Weeks
Working Hours per Year	2080	Hours x Weeks
Number of Technicians	3.8	Technicians
Total Possible Tech Hours per Year	7904	Possible Hours per Day x Days per Year x Techs
Service Bay Productivity (%)	85.0	SBT (%)
Divided by:		
Actual Tech Hours per Year x SBT (%)	6718	Actual Tech Hours per Year x SBT (%)
CODB per Hour x SBT (%)	$45.63	CODB per Hour Factored x SBT (%)

From the model, you can see that our actual CODB per technician hour at 85 percent is $45.63. That is before we can put one penny in our own pocket-for ourselves, for reinvestment, for owner's equity, for additional training, education, more advertising, or retirement!

Just how important is this number? Well, if your CODB is $51.72 as shown in Table 4-7 and your labor rate is $50.00 you are losing $1.72 on every hour you invoice!

And what about a profit? Shouldn't there be some kind of a return on all the time and money you've invested? If you are Oreo pricing, there may be no profit—no return on investment.

WHERE DO WE GO FROM HERE? PERSONAL PROFIT DEMAND

What do we do with this CODB number now that we know how to calculate it?

Using 85 percent service bay productivity as our guide and $45.63 as our CODB per tech hour, the first thing that we have to realize is that $45.63 is our cost. It does not reflect any of the additional demands that being in business places on you as the owner or manager of a business-like the costs associated with the continuing education of your technicians, a figure that experts believe may grow to between 2 percent and 5 percent of your gross sales or more. Sometimes demands are unforeseen-such as a new roof or a fresh coat of paint. Nor does it reflect the additional revenue required to grow your business through marketing and advertising or the investment you will have to make in order to remain current with the demands technology will make on you and your business. Putting all of these financial pressures and responsibilities aside, where is your money? Where is the money you will need for retirement? Where is your return on investment?

It isn't part of the $45.63 unless you consciously include it! Even though your salary might be included in officer's salaries on our original calculations, that $45.63 is just your wages! It doesn't reflect any reward for all the risks you take every day.

The question before us now is: how do you factor a return on your investment into your CODB calculations? Before determining what you want, you should start by figuring out how much money you need—need for yourself, need for rent or a mortgage, need for repairs or improvements on your house or apartment, need for clothing, insurance, entertainment, and for anything and everything else you are likely to need to keep body and soul together. The following is just a partial list of some of the things that might be included. Table 4-9 is a template of sorts that can help you understand what kind of a *profit demand* you will require.

It is probably as complete as you will need it to be and simple enough to flesh out without the help of your CPA. Just remember, if there is something in your world—an expense or financial demand that does not appear here—include it.

Today a good technician should be making anywhere from $35,000 to $65,000 per year. It is not unreasonable to suggest that the shop owner should make at least that much plus a return on investment that reflects the risks, aggravation, anxiety, and commitment involved in any kind of entrepreneurship! Once you have that number, it must be factored into your CODB calculations and added to your CODB per tech hour factored for service bay productivity.

TABLE 4-9 PERSONAL PROFIT DEMAND TEMPLATE

Personal Profit Demand		
Living Expenses (monthly)		
Housing: Rent or Mortgage		
Housing: Maintenance		
Housing: Utilities		
Food		
Personal Needs		
Clothing		
Life and Other Insurances		
Transportation		
Medical		
Education/Recreation		
Gifts		
Savings/Loans		
Miscellaneous		
Total Living Expense		
Business Loan		
Monthly Payments		
Other Loans		
Estimated Tax Liability		
(Federal State, Local, and FICA) Annual		
Monthly Personal Profit Demand		
Total Annual Personal Profit Demand (x 12)		

CODB AND YOUR LABOR RATE

Success in business is based upon a number of things, not the least of which is profits. Profits arise when you sell something, in this case labor, for more than you pay for it. Earlier in this guide, we stated that you must realize an overall average GPM on parts, one of the critical KPIs you must monitor and manage diligently, of 50 percent or more. That takes care of the parts we buy and sell, but what do we do about the service or labor side of the business?

There are a number of different ways we can establish a labor rate without having to resort to Oreo pricing. One is based upon the CODB calculations we just developed factored for service bay productivity and the other is based upon the cost of that labor alone.

Since most of the expenses are included in this first calculation, we are looking for what I would call a net margin of profit on labor. It is a net margin because all of the expenses involved in running the business, with the exception of cost of goods sold, are already factored in. That means all the benefits and taxes are already taken care of, including sick days, holidays, and vacations. All that remains is to plug in our personal profit demand, something for the risk we accept today and a little bit to act as the reward we will want to enjoy in the future, and then run the numbers.

Before you ask, a part of that personal profit demand could be satisfied by the profits we realize on parts, but realistically it probably will not be enough by itself. Consequently, we must margin up our CODB per hour factored for service bay productivity to reflect the additional profit we will need to insure the kind of future we want for ourselves and all the other important people dependent upon the business for survival and success. The only question that remains is how much-10 percent? 20 percent? 30 percent? What is enough? Before we deal with that, however, we should insure that our labor rate is at least equal to our CODB.

Let's see how we would approach this exercise if Above Average Automotive was really ours.

We know that Above Average has a posted labor rate of $50.00 per hour (not very much above average). We also know that with 3.8 technicians working at Above Average Automotive we have the potential to bill a total of 7,904 hours.

3.8 Technicians x 40 Hours Per Week x 52 Weeks Per Year = 7,904 Possible Tech Hours Per Year

That is, 7,904 possible technician hours, if we billed every hour we could.

That didn't happen. We only billed about 63.5 percent of the hours available, realizing labor sales of $249,451 and total sales of $419,047. With a service bay productivity of 63.5 percent, our CODB was actually higher than our posted labor rate: a CODB per hour of $61.08 factored for service bay productivity, versus a posted labor rate of $50.00 per hour. In essence, it is costing Above Average Automotive $11.08 for every hour that is billed!

How does Average Automotive stay in business? Let's see if we can answer that question. Since we already know that with labor sales of $249,451 and a CODB of $306,576, we were not able to cover our CODB, so the first place to look would be in parts sales. Our parts sales were $155,505 and our cost of goods sold was $118,145. That leaves us with a gross profit margin on parts of almost 25 percent and a gross profit in dollars of $37,360. In other words, we can actually say that the only profit Above Average Automotive enjoyed was derived from the sale of parts and accessories.

What happens to all of our calculations if we change our labor rate to reflect our CODB? If we divide our labor sales by our labor rate, we're going to see that we were only paid for 5,019 of the 7,904 hours we could have sold. If we did nothing other than increase our labor rate by about 22 percent, or the $11.08 per hour we are short, what would happen to our numbers? Our labor sales would jump to $306,560. Our total sales would jump to $476,156. And our net income would increase exponentially!

That's what you can do by increasing your labor rate until it is at least equal to your CODB. What would happen if your labor rate was 10 percent higher than your CODB—if you increased your labor rate from $61.08 to $67.18, 10 percent above your CODB? $67.18 times 5,019 would generate labor sales of $337,176, total sales of $506,732 and a net income of $90,528. Not bad, considering we did nothing more than increase revenue without increasing cost.

But we don't have to raise our labor rate to the point of discomfort. We can increase our productivity and have the same desired result. A 10 percent increase in service bay productivity will have the following result—labor sales would increase to $337,271, and parts sales would follow that increase to $171,055. Based upon a fairly constant labor mix, the relationship of labor and parts sales to total sales, it is not unreasonable to assume that our parts sales would increase proportionately with the increased number of labor hours billed.

We know that if our CODB for the year (without cost of goods sold) is $306,576, the per hour figure factored by service bay productivity at 63.5 percent is $61.08. We know the shop is presently doing an annual sales volume of approximately $476,156 based upon the figures we have. We can project that volume by using the primary CODB model in Table 4-10—this shows a shop that is 85 percent productive with annual expenses of approximately $306,576 excluding cost of goods sold and the labor mix percentage that appears on the P&L statement in Chapter 2 (see Figure 2-2).

TABLE 4-10 CODB PER TECHNICIAN HOUR FACTORED BY 85 PERCENT SERVICE BAY PRODUCTIVITY

CODB for 12 Months	$306,576	
Days Open per Week	5	Working Days per Week
Hours per Day	8	Hours per Day
Hours per Week	40	Hours x Days
Weeks Open per Year	52	Working Weeks
Working Hours per Year	2080	Hours x Weeks
Number of Technicians	3.8	Technicians
Total Possible Tech Hours per Year	7904	Poss Hrs per Day x Days per Year x Techs
Service Bay Productivity (%)	85.0	SBT (%)
Divided by:		
Actual Tech Hours per Year x SBT (%)	6718	Actual Tech Hours per Year x SBT (%)
CODB per Hour x SBT (%)	45.63	CODB per Hour Factored x SBT (%)

How much do you margin-up your CODB and how do you do that? Your investment in a good technician will be approximately $800 to $1,200 per week. That's $20 to $30 per hour without benefits—adding 30 percent to that number for taxes, benefits, miscellaneous, etc., our technician cost is about $26 to $39 per hour. If you are supposed to realize a gross profit on labor of 60 percent, that means you will be looking at labor rates of $65 to $97.50 per hour before you start factoring for productivity!

If you want to improve your current reality by finding and keeping a better technical staff, you will have to margin your CODB up to compensate for the difference between what you've got now and what you will need for the future.

WHAT ABOUT INVENTORY?

What about it? We implied earlier that inventory was not a concern because it was a recovered cost. In other words, the money spent on parts and accessories purchased for resale is a *pass-through* cost—parts and accessories are only passing through the business on their way from the manufacturer, warehouse, or jobber to the ultimate consumer, the vehicle owner. That's true—but somehow, some way, some*one* is going to have to come up with the money to buy that stuff.

The money you spend on inventory comes out of the money that your business earns. In order to be successful, you must realize a profit on the parts and accessories you sell just as you must realize a profit on the labor and time you sell. Once again, the issue is one of return on investment, an issue of how well or how hard that money invested in parts and accessories will be working for you. It is also a matter of using what we now know about our CODB calculations to help us generate the dollars necessary to fund the purchase of parts and accessories.

It is our responsibility to insure the highest rates of return possible on any of the investments we make, and the money invested in inventory is no different. This capital could be invested differently and if it was, we would have to ask ourselves what would it return? Whatever that percentage might be, it should be factored into your hourly rate as should future needs for tools and equipment.

With the increasing costs of delivering quality automotive service today, it probably wouldn't be unreasonable to demand the same margin of profit on our labor as we do on our parts, and we have already established a goal there.

If we were to apply the same formula we have been using to our CODB calculations we would divide our $45.63 by 0.6 (to get 40 percent) and come up with a figure of $76.05 per hour! Compared to what we are used to seeing in this industry that isn't a bad place to start.

Can you get it? I don't know and neither do you, not until you try. That number is low in many parts of the country, but it is unthinkably high in too many others.

ALTERNATIVES

For years the experts have been telling us that there are all kinds of different formulas available to help you set your labor prices. In fact, I just read an article that suggested two different ways to accomplish that goal. One was very similar to what we have just done, while the other was perhaps, as was suggested in the article itself, *oversimplified*. The *oversimplified* method (which appeared in a manual published in 1946) suggested that all you needed to do was multiply your technician's salary by 2.5 and you would achieve a gross profit on labor of 60 percent (unloaded)—an accurate and possibly adequate number.

Let's see what happens when we try to do that, but only after we modify the formula slightly to fit our needs. Our original spreadsheet allowed us to look at all of our expenses, a number of them related to payroll. If we took those numbers and then broke them down the same way we broke down the CODB numbers, how much more accurate would it be?

Let's see... Let's say your technicians are being paid $20.00 per hour. And let's say that your financial responsibilities, such as benefits, tax liabilities, contributions, and insurance, add up to an additional 30 percent of that figure (which it will). The total cost of the technician per hour should be right around $26.00. Applying the 2.5 x technician's salary formula above would leave you with 2.5 x $26.00 or $65.00—significantly more than our $45.63 CODB per hour, but not necessarily enough to cover all of the other very necessary expenses you and I will need to recover in order to remain financially strong.

But what would happen if we modified the formula to reflect the changes both in our industry and in the economy that have occurred over the past 50 years? What would happen if we changed the formula to... let's say... three times technician's wages including perks and benefits? We would then have an hourly figure of $78.00, still a little low but clearly better than $45.00 or $65.00 or anything else we might come up with if we depended upon Oreo pricing alone.

We can be even more analytic than that if we want to. We can create a template just like the one used by the Automotive Training Institute, seen in Table 4-11, to help us figure out exactly what our labor rate really *should* be.

TABLE 4-11 LABOR RATE DEVELOPMENT TOOL

Labor Rate Development Tool

	$13.50	$30.00
Technician Pay per Hour *(Obtain this from your P&L shop labor cost ÷ Tech then ÷ hours on the floor)*	$13.50 0	$30.00 0
FICA, FUTA, SUTA ⇒ *(Approx. 11.5 % x hourly rate)*	$1.55	$3.45

Dollar value of benefits you provide:

	Annually		Monthly		Weekly		Hourly	
HEALTH INS		$2,059.20		$171.60		$39.60		$0.99
LIFE INS		$93.60		$7.80		$1.80		$0.05
DISABILITY INS		$374.40		$31.20		$7.20		$0.18
UNIFORMS		$421.20		$35.10		$8.10		$0.20
VACATION	$1,216.80	$1,521.00	$101.40	$126.75	$23.40	$29.25	$0.59	$0.73
HOLIDAYS	$608.00	$760.00	$50.67	$63.33	$11.69	$14.62	$0.29	$0.37
LUNCHS								
PERSONAL DAYS	$608.00	$760.00	$50.67	$63.33	$11.69	$14.62	$0.29	$0.37

				$16.22	$36.33
Actual Technician Cost per Hour ⇒				$16.22	$36.33
Desired GPM = 60 % X 2.5 = 60 %				$40.56	$90.82
Adjusted for Service Bay Productivity by Tech 100 % 60 %				$67.59	$90.82

If you follow the numbers across and down in the labor rate development tool in Table 4-11, you will find a simple, easy, and most importantly, an accurate alternative to Oreo pricing. In the upper right-hand corner of the diagram you can see that we are looking at technician's wages that can range from a low of $13.50 to a high of $30.00. That would take you from an entry-level, apprentice technician to an ASE Master Tech with your L1. To the base salary you would add approximately 11.5 percent for all the state, local and federal taxes required of you as an employer and then all the benefits that you see itemized along the left side of the development tool, such as health insurance, life insurance, disability insurance, vacation, sick or personal days, uniforms, and any other benefits of employment you choose to provide. These benefits are presented as a yearly expense first, then divided by twelve and presented as a monthly expense, then broken down by the week and finally by the hour.

As you can see, with all the *adds* your $13.50 base salary has jumped to $16.22 and the $30.00 base cost for an A tech has jumped to $36.33. That's a quick 20 percent increase.

Our next step would be to increase our base numbers once again to reflect our desired GPM of 60 percent, and we can do that by multiplying that modified base by 2.5 or dividing by 0.4. That will take our labor rate calculation to $40.56 and $90.82. The next and final step in this process would be to factor for technician productivity. In the case of an entry-level technician, it wouldn't be reasonable to assume 100 percent productivity. An apprentice or beginner has an awful lot of information to process, an awful lot to assimilate. Realistically, assuming a productivity level of 60 percent for a C tech would not be that far off the mark, so we'll multiply by 2.5 or divide our $40.56 by 0.4 once more to get an entry-level technician's labor rate (lube jobs, basic brake work, etc.) of $67.59.

As far as our A tech is concerned, at $90.82 per hour that technician had better be cranking out 100 percent or more! So there should be no further adjustment necessary there.

The bottom line is simple. When all is said and done, there just isn't any reason to run with the pack when you can be the market leader in your area. There are more than enough alternatives presented here to prove that Oreo pricing is absolutely unnecessary.

WHAT ABOUT REALITY?

"The economic conditions in my area just won't support a $70.00 or $80.00 an hour labor rate. That's just plain impossible!" you say?

Well, in many instances that is a very real consideration. You see, if you and I charged only what our CODB numbers suggested we should, factored for productivity and the need to satisfy our personal demands, the hourly rates would still fluctuate madly because not everyone is in the same place in the life of their business. Exactly what does that mean? Let me show you.

Many years ago, while presenting a seminar in Texas, I met a garage owner who was particularly proud of the fact that he was still able to charge only $22.00 per hour—and still make a profit! He could do that because he had long since paid off just about every debt he had ever incurred and hadn't invested a nickel in his business for years! He didn't have the same commitment to hi-tech equipment, education, or training that he might have had earlier in his career, and he certainly didn't have the same economic burden to carry. Consequently, he really could set his prices significantly below the competition and still realize a profit. But was that the right thing to do—right for him, right for his market, right for the industry? Was it fair to his competitors? And ultimately, was it really fair to his family or to the individual who would have to buy the place from him in order to fund his retirement?

The answer is a resounding NO!

Just because you have been in one place forever, just because your equipment is all paid for, just because your cost of doing business is legitimately lower than your neighbors' is no reason to allow your prices to hover just above the floor!

What happens when you go to sell out? Who will be able to make the improvements that will be necessary? How will whoever buys the place be able to pay you, particularly if he or she has to raise the prices to do it? What are your customers going to think when the first thing the new owner does is raise the labor rates?

Forget about all of that. Why aren't you entitled to make a better return on your investment after a lifetime of effort?

My philosophy is simple and I'm not at all ashamed of it. All of us have the responsibility to push the envelope as far as it will go with regard to price! Why? Because our prices have been so low for so long it is our right! For almost one hundred years we have been pricing our goods and services far below the other skilled trades, far below what they were worth—in essence, subsidizing the repairs of our customer's cars all the while.

If your CODB suggests to you that you can set your prices far below your neighbors' and still make a profit (however small), why not set your prices right in line with theirs and make a still larger profit? Why not offer more and better services, value-added services, and charge more, making an even better return on your investment? Is it really such a crime if you can deliver an excellent quality product and still satisfy your customer's needs?

And what do you do if your CODB demands an hourly labor rate far and away above those of your competitors? What do you do then? Can you really set your prices that far above your neighbors' and still remain in business? If you sit down and make the commitment to go through these calculations, determining exactly what your retail labor price should be, and it is just unrealistically high, too high to even consider posting, what do you do then?

Well, there is only one thing you *can* do and that is compensate! You create efficiencies to improve productivity wherever you can. You price your parts to help create the profit structure you need to survive. And you don't have to offer services that are not profitable! You look at alternative services that you can perform, other profit centers that you can explore, and then you adapt, adjust, and overcome!

Whichever situation you find yourself in—CODB too high or too low—keep your prices at the edge of the envelope, always presenting you with the highest possible returns. Success and failure are in your hands. As the owner or manager, it's your responsibility.

REVERSE BUDGETING

Budgeting is one of those business concepts that has consistently driven me nuts. For years, it just didn't make any sense to me. I understand the need to budget on the basis of what I have seen and experienced, but in the real world it never worked for me—perhaps because I was never really goal-oriented when I was younger. Or at least I didn't realize just how goal-oriented I managed to be without a formal set of written goals in my pocket or in my toolbox. Aside from that, budgeting is a fairly detailed exercise and requires more than a little organization. Details were never my strong suit and the only place I ever showed any organizational aptitude was on the shop floor when I was diagnosing or repairing a car or truck.

Then I was introduced to a different kind of budgeting and that has made all the difference in the world. It is a kind of reverse budgeting, and the reason it makes sense to me is that it starts with the desired result and then fills in all the long- and short-term goals and objectives for you. I liked it so much I built a spreadsheet for it and started using it a number of years ago myself. It's not all that difficult, because you and I have already been over most of the calculations already. So finding the information to complete the formula isn't a problem. At least, it shouldn't be.

The spreadsheet appears in two sections; an upper section containing a number of *givens* or *knowns*, and a lower section where all the numbers get crunched. We'll start out with the upper section first.

You start by picking the number you would like to see on the bottom line of your financial statement and then work backwards from there. I have chosen $120,000 as an example in Table 4-12. If we go back to Table 4-2 at the beginning of this chapter, we can see where the numbers come from. On the initial CODB worksheet we showed yearly expenses of $306,576 excluding the cost of goods sold. If we break out direct labor, the cost or wages associated with creating that labor, and then divide the result by the number of days we are open, we have the first number we are looking for:

Daily Operating Expense without Technician Pay

Crunching the numbers would look something like this:

$306,576 - $76.964 = $229,612 ÷ 260 = $883.12

260 is the number of days we are open per year—5 days per week x 52 weeks per year.

Parts purchases can be found in Table 4-1. It is the number associated with Purchases: Parts in the Direct Cost section of the financial statement—$118,145.

After parts purchases come the following givens: percent profit on parts/accessories, labor cost as a percentage (of your labor rate), number of tech hours open, percent productivity, and number of technicians. Since we are reverse budgeting here, trying to figure out what we need to do in order to reach our desired goal of $120,000, we don't necessarily have to use our own numbers. We can project or forecast a number of *what if* scenarios by substituting numbers we would like to achieve rather than using numbers that we already have. As an example, let's say that your percentage of profit on parts/accessories is 38 percent, but you would like to know what would happen to your numbers if you were able to increase that percentage to 40 or 45 percent. You can do that here.

In Tables 4-1 and 4-2, Above Average Automotive isn't doing so well. In fact, Above Average's performance isn't above average at all! With parts sales of $155,505 and a parts cost of $118,145, Above Average enjoyed a gross profit of only $37,360 or a GPM of about 24 percent. We could certainly include that 24 percent as our percentage of profit on parts/accessories to see what we have to do if we are unable to increase our margins based upon what we now know is necessary,

TABLE 4-12   REVERSE BUDGETING

Reverse Budgeting		
$120,000.00	Desired Profit	
$883.12	Daily Operating Expenses without Tech Pay	
260	Number of Days Open	
$118,145.00	Parts Purchases	
40%	Percent of Profit on Part/Accessories	
30%	Labor Cost as a Percentage	
7,904	Number of Tech Hours Open	
85%	Percent Productivity	
3.8	Number of Technicians	
$120,000	Profit	
$229,611	Expenses without Tech Pay	
$118,145	Part Cost	
$467,756	Total Outflow	
$196,908	Part Sales @	40 %
$270,848	Labor Profit Needed	
$386,926	Labor Sales	
$57.59	Labor Sales per Hour	
$15.16	Labor $ per Hours per Technician	
$583,834	Annual Sales	
$48,653	Monthly Sales	
$12,803	Monthly Sales per Tech	
$591	Daily Sales per Tech	

or we can assume that we will be at least partially successful in our attempts to increase margins and that number will increase to the 38 percent figure we used above—or something even higher.

The same can be said for labor cost as a percentage. We can use our actual labor cost as a percentage or move that number higher or lower depending on where we are in our own business. If our labor cost is too high, we can see what would happen if we could somehow lower it (although that might not be the most prudent thing you could do unless you have some terribly unproductive, inefficient technicians working with you!). If the labor cost is too low, which is much more likely to be the case in our industry today, you could actually see the impact of investing in an A tech.

The number of tech hours open is easy. That is the number of technicians we have working (a number that will find its way to the last box in this section) times the number of days times the number of hours.

3.8 Technicians x 260 working days x 8 hours per day = 7,904

That is the same number you will see in the middle of Table 4-5.

We have already discussed service bay productivity and the formulas associated with calculating it. We can take the same principles and put them to use here. We know that Above Average Automotive's labor sales were $249,451. We know that our labor rate is $50.00. If we divide our labor sales by our labor rate, we have an approximate idea of just how many hours we invoiced or billed.

$249,451 Labor Sales ÷ $50 per hour Labor Rate = 4,989 billed or invoiced hours

If we multiply the number of technicians we have times the days and hours they are at work, we have the total available hours, or 7,904.

Actual Hours ÷ Available Hours = Productivity %

4,989 ÷ 7,904 = 63 %

Do we use the 63 percent figure or do we substitute a number representing a higher degree of productivity based upon our new knowledge of its importance?

The lower section in Table 4-12 starts out with our desired profit and then drops immediately to daily expenses without technician pay. That number is calculated by taking the number of days you are open and multiplying it by your daily expense without tech pay. In this case it is:

260 days x $883.12 = $229,611

Following that, we will be inserting our part cost and then totaling those three items together.

$229,611 + $118,145 + $120,000 = $467,756 Total Cost or Outflow

The next two calculations might seem a little more difficult, but they really aren't. We have the total of all of our expenses without technician pay, the cost associated with the parts we have purchased, and our desired bottom line profit (which could be over and above our salary!). We will add to that what our parts sales should be factored for the GPM we have chosen. In this case it will be $118,145 divided by 0.6 or multiplied by 1.67 (you can find the appropriate chart in Figure 3.1 at the beginning of Chapter 3: Markup and Margin), which equals $196,908.

If we then subtract that $196,908 from our total outflow or cost, what will be left is the labor profit required to meet that demand or the labor profit needed—in this case $270,848. If we know what profit we need and we know what our desired GPM on labor is, we can figure out what our sales must be, very much the same way we calculate margin.

Here we know that we need $270,848 in labor profit and we also know that we want our cost of labor to remain at 30 percent of sales. To find out what those sales should be, we divide the number we know, $270,848, by 70 percent or 0.7, the amount of desired gross profit.

$270,848 ÷ 0.7 = $386,926

So with the information we already have and the desired profit we have suggested is necessary, we now know that Above Average Automotive must generate annual sales of $583,834 in order to create $120,000 of desired profit. That translates to $48,653 per month, $12,803 per month per tech, $591 per day per tech, $57.59 in labor sales per hour.

Just remember that if we change anything, we change everything.

RECOVERING EXPENSES

There is another useful tool that has changed the way I look at my business forever, particularly my expenses. From almost my first visit with an accountant, I've been taught that if you really want to increase bottom line profits and make more money, there are only three ways you can do that. You can decrease expenses, you can increase gross sales or you can increase gross profit. That's it! There isn't anything else you can do, no other way to do it. And for years I believed it. Then someone suggested there was another way to make more money, another way to generate profit, and that was by recovering expenses. At first, I didn't understand what he was trying to tell me. But after a while it began to make sense.

We're taught to cut or reduce expenses almost from the moment we go into business. But doing that is almost always a one-time fix. If you continually go back to your expense account trying to trim away the fat every time you want to see a few extra dollars in the bank, you will ultimately run out of fat and cut away muscle and bone! There is another way. The only problem is that the other way requires that you stop thinking about your expenses as expenses and start thinking about them as investments.

If you think about purchasing something for your business, almost anything, in terms of its cost and its cost alone, your natural response will be to limit that cost or expense as much as possible. However, if you think of that cost or expense as an investment the whole thought process begins to change.

With an investment, all we are really interested in is how much it will cost initially, how long it will take to get some kind of a return on that investment, and how much of a return that is likely to be.

Consequently, what I am suggesting here is that we change our paradigm a little and start thinking about our expenses as if they were investments.

As an example, I'd like to consider the investment you and I will be forced to make in order to keep our technicians educated, trained, and technically competent. I have chosen training expense because it is perhaps the single most onerous expense there is for most shop owners or managers. Why? Because they are convinced that money invested in training is wasted, especially if that technician decides to leave the shop and take that education with her.

So let's see what we can do when we change the training paradigm a bit and start thinking about training as an investment. The first question is: How much will it cost? I'm going to make this example particularly ugly! I will suggest that we are going to have to send a technician off-site to an out-of-state training center somewhere far away from you. That will entail a number of additional ancillary expenses like travel, lodging, meals, and more. For our purposes, the training that we choose to access is $1,000 for four-and-a-half days of automotive service education, which will be reflected in Table 4-13. Because the training is only available out of state, there will be the additional expense of $350 in airfare to get the technician there. Lodging will be $60 per night, or $300 for the five nights the technician is there. A reasonable per diem for food might be $35 per day, or $175.

We wouldn't be sending just anyone out of town for training, so let's assume that this is one of our lead technicians and her guarantee is $800 per week. Since the technician is going away on company business, she has to be paid, so we'll drop that into the equation below as well. The next issue we will have to deal with is lost revenue, the money we won't make with one technician out of the service bay. If we use the numbers we have generated in Table 4-12, we know that to generate the kind of volume that we need to achieve our desired profit, each technician has to do $12,803 in monthly sales. Divide that by 4.3, the average number of weeks in a month, and you have lost revenue of almost $3,000 ($2,977).

TABLE 4-13   RECOVERING EXPENSES

Recovering Expenses: Training	
Cost of Training	$1,000.00
Travel	$350.00
Lodging	$300.00
Meals	$175.00
Salary/Guarantee	$800.00
Lost Revenue	$3,000.00
Total Expense	$5,625.00
2-Year Return on Investment (ROI)	
Hours per Day	8
Days per Year	260
Hours per Year	2,080
Number of Years	2
Total Number of Hours	4,160
Increase per Hour	$1.35

The total of all the expenses involved in sending this particular technician for training is $5,625. That is a substantial investment by anyone's standards and not something any garage owner or manager would take lightly. If it was considered purely as an expense, it is unlikely it would ever happen. However, if you start thinking about it as an investment, a number of dynamics come into play that change the whole nature of the relationship you have with those numbers.

As an investment, we said we need to know how much capital (money) will have to be invested, how long it would take to achieve a return on that investment, and how large a return it might be. We already know the first amount—it's $5,625. So the only two things remaining are how long and how much.

The how long is up to you. But with a $5,625 investment, I would say two years would not be unreasonable. So we can drop that into our equation below, and that's when the fun begins. We know that our investment will take two years to return some kind of a dividend. We also know that our technician will be working eight hours per day, 260 days per year, for a total of 2,080 hours per year or 4,160 for the life of the investment.

If we divide the $5,625 we have invested by the 4,160 hours we have allowed to achieve some kind of a return on that investment, all we have to do to recover that expense is add $1.35 per hour to every hour the technician invoices the minute she returns from training!

In my opinion, this is the gift that keeps on giving! The ultimate investment! Why? Because it benefits the shop, the technician, and you as the shop owner. The technician has new skills and abilities that will hopefully increase her value. Those skills and abilities can be shared with the other technicians in the shop, reducing the possibility they might have to get on an airplane and go off-site. And the technician is more secure and confident knowing that you have made that kind of an investment in her future.

It is the *gift that keeps on giving* for another reason; it is unlikely that you will remove that $1.35 increase after the two years are over. More than likely, it will remain a part of that technician's labor rate, and that will help generate the revenue for the next class. And, perhaps best of all, despite the fact that you and I might agree that this expense has been somehow magically transformed into an investment, your accountant and the IRS still agree it is an expense and therefore can be written off, limiting your tax liability—not such a bad investment after all!

SUMMARY

Most independent shop owners rarely take the time or dedicate the energy to actually set their prices based upon anything more than what the guy down the street, up the block, or around the corner is doing. This can be economic suicide.

We can't depend upon anyone else to know what it costs each of us to operate our businesses. Trying to create value by setting our prices lower than other service providers in our market area will not insure success. The only intelligent way to set your prices is by knowing and understanding what your costs are and then factoring those costs for service bay productivity. It also helps to know how much money you want to take home, not just how much you need.

Creating a labor rate based upon CODB or upon the total cost of that labor including benefits and all the applicable taxes is the only thing that will make sense for you or your business.

Using reverse budgeting to recognize what the goal is can be an incredibly useful tool when it comes to determining just how much volume you will have to do to succeed, just as forcing yourself to look at your expenses as investments can change the way you approach such critical investments as training or recapitalizing your equipment.

Think about how you and perhaps everyone else you know has calculated their labor rates. Think about whether or not any conscious thought went into the process at all. Did they Oreo price and just pick a number in the safe, creamy center of the cookie? Or did they actually do the math? Did you?

In the chart on the next page, write down the steps you will have to go through to create a labor rate that works, a labor rate based upon something more than what the guy across the street is charging. What will you have to do to insure that your labor rate is correct and adequate?

Thoughts

Actions

Results

CHAPTER 5

Getting From A To B

INTRODUCTION

This chapter introduces you to ten of the critical KPIs necessary to effectively monitor and manage your automotive service business:

- GPM on labor

- GPM on parts

- Car count

- Average invoice or average repair order

- Labor mix

- Labor content per job

- Technician efficiency

- Service bay productivity

- Total sales per bay

- Total sales per technician

It will also show you how to use these critical KPIs to move your business toward success.

GETTING STARTED

The start of a journey like this begins by firmly establishing where you are in your business at a particular moment in time. Knowing where you want to be is irrelevant if you have no idea where you will start the journey. You do that by identifying the processes in your business critical to success and then breaking those processes down into metrics that will accurately allow you to measure performance. The results of such an exercise can be called many things, but we are going to call them KPIs.

If you understand what the key processes are and know what numbers it will take to reach the targets you are shooting for, you can quickly identify just how well (or not so well) you are doing.

BENCHMARKING: KEY PERFORMANCE INDICATORS (KPIs)

There is probably no limit to the number of things you can measure in any business, ranging from basic productivity to cost per square foot of service bay floor space. I could write a book on this one subject alone, but I'm not sure that would help as much as achieving a basic understanding of the principles we are discussing here and applying what we have learned to the real world. Consequently, we will only be looking at ten of these KPIs. However, the ten we will look at will provide nearly all the information we need to determine the physical health of our business at any given moment and certainly what our business looks like when compared to another business in the same industry.

The ten we are going to look at are GPM on labor, GPM on parts, average invoice, car count, labor mix, labor content per job, labor sales or total sales per bay, labor sales or total sales per technician, efficiency, and productivity.

Just a note about what we are *not* going to look at. We are not going to look at net profit for an automotive service business because there are too many external influences that can alter that number dramatically, such as whether the business is a sole proprietorship (the business is owned by a single entity), a partnership, or a corporation. We are not even going to consider gross profit on the entire business other than to try to understand how that number is impacted by these other ten KPIs—because *how* and *where* you report your financial information on the income statement can have a significant impact on gross profit as well. We will, however, spend a fair amount of time on capturing these ten KPIs, because without knowing what they are and what they are telling you about your business, you are piloting an airplane that has no windows without being instrument rated!

We have said repeatedly throughout this series that if you can't measure it, you can't manage it! Now we're going to take a look at what needs measuring.

GPM on Labor

Generally speaking, gross profit and GPM on labor can be defined as the hourly labor rate less the cost of direct labor—the technician's wages or salary per hour associated with making those sales. This can be reported *loaded* (reflecting benefits and ancillary costs like health insurance, sick or personal days, vacations, uniforms, and a host of other expenses) or it can be reported *unloaded* (without the load associated with those benefits). My suggestion would be to move all the direct costs associated with producing labor sales to the top of your financial statement (covered in another book in this series) where you, your bookkeeper, or accountant generally reports cost of goods sold—the cost of the parts and accessories you purchase for resale.

The cost of direct labor would include all the costs we just mentioned: sick or personal days, uniforms, vacation pay, and holiday pay, as well as life, health, and disability insurance. This will give you a far more accurate picture of what your overall gross profit and GPM are because it will more accurately represent the costs involved in producing those labor sales.

Before you get confused, the difference between gross or total sales and those sales minus the costs associated with them (sales minus cost of sales equals profits) can be reported in two ways. Gross profit is almost always presented by your bookkeeper or accountant in *dollars*, while GPMs are almost always presented in *percentages*.

A formula reflecting the gross profit on labor would look something like this:

Hourly Labor Rate - Cost of Direct Labor Per Hour = Gross Profit on Labor Per Hour

Let's give this formula life by substituting actual numbers for the descriptions. Let's say your labor rate per hour is $65.00. You pay your technician an hourly rate of $17.20, unloaded. The load of taxes, benefits, and associated expenses can drive your total direct cost of labor up anywhere from 20 to 30 percent higher than the technician's hourly rate unloaded. Splitting the difference at a 25 percent load would add $4.30 in associated cost to the technician's hourly rate, or a total direct cost of labor of $21.50.

Knowing your hourly labor rate and your cost of direct labor allows you to subtract one from the other—$21.50 from $65.00—leaving you a gross profit of $43.50 and a GPM of 67 percent.

As a general rule of thumb, your GPM on labor should range somewhere near 60 percent with the benefits and taxes loaded in and about 70 percent with them removed.

Rule of Thumb

Gross profit on labor is approximately 70 percent with core benefits, 60 percent without.

Gross profit calculation:

$65.00 (labor rate) – $21.50 (labor cost per hour) = $42.07 (gross profit on labor)

GPM calculation

Divide gross profit by labor rate

Gross Profit ÷ Hourly Labor Rate = Gross Profit (expressed as a percentage)

$43.50 ÷ $65.00 = 66.9%

Your numbers $__.__ ÷ $__.__ = ___%

GPM on Parts

You calculate gross profit and GPM on the parts and accessories you sell using the same principles you use when calculating gross profit and GPM on labor. To do this, you start with the parts price (the price the customer will pay for that part or accessory), and then subtract the cost of goods sold or the *acquisition cost* (your cost to acquire that part). The result is your gross profit on parts.

Stated another way:

Selling Price (parts price) - Acquisition Cost (cost of goods sold) = Gross Profit on Parts

Example: A part that will be sold for $60.00 cost you $36.00 to acquire (or purchase) from your local parts supplier. We give this formula life once again by substituting actual numbers for the equation above. To calculate gross profit, you subtract the cost of the goods you are selling from the selling price and what is left is your gross profit: $60.00 - $36.00 = $24.00.

Rule of Thumb

Gross profit on parts should average 45 to 55 percent or more, depending upon the selling price. (In some instances, higher priced items may require lower margins because consumers seem to be more price sensitive when they are reaching deeper into their savings.)

Gross profit calculation:

$60.00 (selling price) – $36.00 (acquisition cost) = $24.00 (gross profit on parts)

GPM calculation

Divide gross profit by selling price

Gross Profit ÷ Selling Price = Gross Profit (expressed as a percentage)

$24.00 ÷ $60.00 = 40.0%

Your numbers $__.__ ÷ $__.__ = ____%

Knowing the gross profit (represented in dollars) will allow you to calculate your GPM on parts, which is also a critical number for you to know and understand. To calculate GPM on parts, you divide your gross profit on parts by the selling price of that part or accessory—$24.00 ÷ $60.00 = 40 percent.

I used a GPM on parts of 40 percent here, *not* because it is the target, but because it is probably the minimum average GPM on parts you can accept and still hope to remain in business. Overall, GPM on parts should average 45 to 55 percent. Low margins will compromise your ability to make a decent living, still invest in your business, and possibly even pay your bills. However, higher margins can often affect sales in a negative way by discouraging the more price sensitive segment of your market from trading with you. For this reason, a number of automotive repair shops utilize a concept known as *matrix pricing* for their parts sales which utilizes higher margins for the lower-priced parts and lower margins for the higher-priced parts.

A parts pricing matrix for items in your inventory or for parts acquired from a local supplier (which are generally purchased at a deeper discount than parts purchased from the local dealership) can look something like the one in Table 5-1.

TABLE 5-1   PARTS MARGIN MATRIX FOR TRADITIONALLY SOURCED PARTS
(JOBBER STORE, SHORT-LINE WD, PARTS HOUSE)

NORMALLY SOURCED PARTS MATRIX		
Parts Cost	**Desired GPM %**	**Divide By**
$ 0.00 – $ 5.00	70.0	0.30
$ 5.01 – $ 10.00	60.0	0.40
$ 10.01 – $ 75.00	55.0	0.45
$ 75.01 – $ 200.00	50.0	0.50
$ 200.01 – $ 500.00	42.5	0.58
$ 500.01 – $ 750.00	35.0	0.65
$ 750.00 – $UP	30.0	0.70

The parts pricing matrix for parts sourced at a local franchised dealer could look like the one in Table 5-2.

TABLE 5-2—PARTS MARGIN MATRIX FOR PARTS SOURCED AT A LOCAL FRANCHISED DEALER

DEALER SOURCED PARTS MATRIX		
Parts Cost	**Desired GPM %**	**Divide By**
$0.00 – $1.00	70.0	0.30
$1.01 – $5.00	65.0	0.35
$5.01 – $25.00	60.0	0.40
$25.01 – $75.00	55.0	0.45
$75.01 – $500.00	42.5	0.58
$500.01 – $750.00	32.0	0.68
$750.01 – $UP	28.0	0.72

The overall philosophy is simple. Parts that fall below a certain dollar threshold aren't parts the consumer is normally sensitive to. The vast majority of the sales that make up your parts mix fall within the category of prices under $100.00. Most consumers are far more sensitive to the price of a $2,300 rebuilt transmission and the profit you are taking on it than they are to the nuts and bolts that cost them $.95 but only cost you $.20.

No matter how you calculate GPM on parts, whether you use the matrix models above or the formulas we've just discussed, you must remember one thing. Adequate parts margins are the key to success in any business, not just ours.

CAR COUNT

Car count is another critical number when it comes to benchmarking your business. Your car count is simply the number of cars that pass through the shop every day, week, month, or year. It includes *all* the cars that find their way in and out of the shop, even the vehicles for which there is no charge. If you use sequentially numbered invoices, you can subtract the lower number (the one that appears on the first invoice for that period) from the larger number (the number that appears on the last invoice or repair order for that period). Then divide that number by the number signifying the appropriate time period.

Car Count = Total Vehicles ÷ Calendar Period (day, month, year)

For example, if the first invoice for the previous month was numbered 23390, and the last invoice for that month was numbered 23588, you would subtract 23390 from 23588 and the total number of invoices—or the *car count*—for that month would be 198. If you were calculating daily car count and that particular month had 22 working days, you would divide 198 by 22 for a daily car count of 9 cars per day. You can see the formula and how the numbers develop in Table 5-3.

TABLE 5-3—CAR COUNT CALCULATION

CAR COUNT			
Start date	July 1, 2002	Invoice Number	23588 *
End date	July 31, 2002	Invoice Number	23390
		Total Invoices	198
		Divided by Working Days	22 **
		Equals Daily Car Count	9

* All invoices, including $0.00 Invoices
** Reflects July 4th Holiday

Car count is important because of the impact it has on scheduling, productivity, and technician efficiency, among other things. When car count is too high, or at least high enough to become overwhelming for the number of technicians available, efficiency and productivity begin to erode. When there are too many cars being forced into the pipeline, the quality of the work performed on the cars coming out of the pipeline will more than likely be compromised. Quality will suffer as accuracy is sacrificed at the altar of expediency.

Higher car counts generally equal lower productivity and a lot of undone service work identified and itemized in the *remarks* or *recommended service* area of the invoice. These lost service opportunities are synonymous with lost profits in a world in which time is one of the most precious commodities any of us may have, and getting the vehicle back to the shop may be far more difficult than getting the work done while the vehicle is already in the shop.

So, the key to a successful car count isn't just working on fewer cars. The key is working on fewer cars, but doing more work on the cars you are working on.

Average Invoice (Average Repair Order)

Of all the KPIs, average invoice is perhaps the most telling. Although it may be the simplest to understand and the easiest to calculate, it can furnish more information about an automotive service business, particularly an independent automotive service business, than any other single KPI.

$$\text{Average Invoice} = \text{Gross Sales} \div \text{Car Count}$$

As you can see in Table 5-4 below, your average invoice is the product of your gross sales for any given period of time divided by your car count. If we were looking at a three-bay service facility with average monthly gross sales of $33,333 (a $400,000 per year enterprise) and a monthly car count of 198 vehicles, the average invoice would be calculated by dividing the $33,333 monthly sales figure by the vehicle count of 198 per month, in which case your average invoice would be $168.35 (very close to the national average).

TABLE 5-4   AVERAGE INVOICE CALCULATIONS

AVERAGE INVOICE	
Average Monthly Sales	$33,333
Divided by Car Count	198
Equals Average Invoice	$168.35

High average invoice usually correlates to lower car counts and higher profits in most shops. The higher the average invoice, the higher the productivity; the higher the productivity, the higher the profits. High productivity and higher profits are the product of thoroughly inspecting the vehicle, analyzing service needs, presenting those needs to the vehicle owner, and asking for the opportunity to perform the service. That translates into more service work per vehicle and results in a technician performing multiple service operations on the same vehicle.

Lower average invoices generally mean higher car counts, and higher car counts almost always indicate a high rate of *turn and burn*, a characteristic of a shop in which cars are blowing in and out of the service bays with no time for thorough inspection, service, or repair. Vehicles are leaving the facility with undone service work that just might find its way out of your operation and to the door of another service provider who understands just how difficult it is for most vehicle owners today to get a vehicle into the shop in the first place.

Labor Mix

Labor mix is the percentage of labor sales to total sales, as compared to the percentage of parts sales to total sales, expressed as a ratio. Sounds kind of confusing, doesn't it? But it isn't really.

Up until very recently, automotive shop owners (particularly independent automotive shop owners) were taught that they should work toward $1.00 worth of parts sales for every $1.00 worth of labor sales. In other words, total sales should be split 50/50 between the sales of parts and labor, achieving a ratio of 1:1.

Once upon a time in this industry, that may have been good enough, but it isn't any more. With automotive service work becoming continually more and more labor intensive, a 50/50 or 1:1 model just won't work any longer. There are too many labor dollars evaporating off the service bay floor.

Looking at one hour's worth of work consistent with the old paradigm of a 1:1 or 50/50 parts/labor split, the following numbers jump out at us-or at least they should. Our cost of direct labor is $22.93. (The direct labor cost has increased to $22.93 to reflect increases in continually rising benefit and health care costs.) That number reflects the load of taxes and benefits. Our gross profit on labor is within the 60 percent GPM we're looking for with the taxes and benefits loaded in (70 percent without). Our gross profit on parts is sitting at 40 percent, which is the minimum we'd like to see as an average GPM for parts sales. And when we look at profit per hour, we see that we received $120.00 for one hour's worth of work: $24.00 gross profit on parts and $37.07 gross profit on labor, with a total gross profit of $61.07, or just about 51 percent.

All in all, that doesn't seem quite so bad-and it isn't, compared to the extremely poor performance of a majority of shops engaged in automotive service and repair. However, it's not as good as the performance we'd like to see, performance that starts to improve the minute we tip the scale and begin to recover some of the lost labor dollars slipping through our fingers. You can see what the traditional model looks like in Table 5-5.

TABLE 5-5   LABOR MIX: THE OLD 50/50 MODEL

LABOR MIX: THE "OLD" MODEL		
50% Labor-50% Parts		
Labor Sales	$60.00	
Gross Profit (%)	62	
Gross Profit	$37.07	
Cost of Labor	$22.93	
Parts Sales	$60.00	
Gross Profit (%)	40	
Gross Profit	$24.00	
Cost of Parts	$36.00	
Total Sale		$120.00
TOTAL GROSS PROFIT		**$61.07**
Gross Profit Margin (%)		50.9

These labor losses are the result of a shift in the industry from low-tech or no-tech to high-tech and high-touch-from removal and replacement (R&R) to inspection, testing, analysis, evaluation, diagnosis, and repair. Being able to ascertain what the problem might be and, perhaps more importantly, what might be causing it has become an increasingly important part of basic shop operations as vehicle sophistication and technology have blossomed. Determining what needs to be done has become just as important as knowing how to perform the appropriate repair. Vehicles require two or three road tests to identify symptoms and then confirm proper correction, and all of this takes time that can be translated into labor hours that aren't always being charged out on the customer's repair order.

To be successful in the automotive service environment today requires a parts/labor split that more closely approaches 60/40 than it does 50/50, with labor sales moving as close to 60 percent of total sales as possible. When this begins to happen, the results are almost instantaneous and very dramatic, as you can see from Table 5-6.

TABLE 5-6   LABOR MIX: THE NEW MODEL

LABOR MIX: THE "NEW" MODEL		
55% Labor–45% Parts		
Labor Sales	$74.00	
Gross Profit (%)	69	
Gross Profit	$51.07	
Cost of Labor	$22.93	
Parts Sales	$60.00	
Gross Profit (%)	40	
Gross Profit	$24.00	
Cost of Parts	$36.00	
Total Sale		$134.00
TOTAL GROSS PROFIT		**$75.07**
Gross Profit Margin (%)		56.0
"New" Model Gross Profit		$75.07
"Old" Model Gross Profit		$61.07
Difference		$14.00

You might notice a couple of differences between the *old* model in Table 5-5 and the *new* model in Table 5-6 almost immediately. First of all, you *should* have noticed that the labor rate increased from $60.00 for one hour's worth of work to $74.00. It had to. No one wants you to sell fewer parts in the same hour's worth of work. We would, however, like to see you recover some of the lost labor time we've been talking about, and the easiest way to do that is to charge more per hour for every hour's worth of work you charge for! If our parts sales remain the same, and that number becomes anything less than 50 percent of total sales, our percentage of labor sales for the same period must go up. Consequently, the increase is reflected in Table 5-6. Our direct cost of labor has increased to reflect that fact with a higher labor rate we could conceivably invest more money in direct labor, and as a consequence attract a somewhat better qualified technician.

Note that with all things remaining virtually the same, with the one exception of an increase in labor sales, our gross profit jumped to $75.07, or $14.00 more than the same hour's worth of work in the old model.

There are more advantages to adopting this kind of paradigm than can be discussed here. However, don't be fooled. The advantages we look at here can and will make a powerful and positive difference in your business. First of all, our profit per hour increased by $14.00—that is almost a 23 percent bump in gross profit for the same hour's worth of investment. If you can invoice eight hours in an eight-hour day and sell the appropriate amount of parts, you could realize an increase in revenue of $112 per day, and that is with a 55/45 split, not the targeted 60/40 split.

This bump occurs because the GPM on labor is substantially higher than the overall GPM on parts, and any increase in labor sales is reflected on the bottom line almost instantly.

You really don't have to calculate this ratio on your own. The numbers already exist, and if you have a computerized automotive shop management system, a bookkeeper, or an accountant, finding the numbers is really pretty straightforward. In most financial reporting systems, each component of the income statement is reported as a percentage of total sales. Consequently, all you need to do is look at that statement, an example of which you can see in Table 5-7, and you will begin to see the correlation we are looking for.

TABLE 5-7   LABOR MIX DISPLAYED AS A PERCENTAGE OF TOTAL SALES ON THE INCOME STATEMENT

AVERAGE AUTOMOTIVE, INC. **STATEMENT OF REVENUE & EXPENSE** Period Ended – August 31, 1999				
	One Month Ended			
	31-Aug-99		31-Aug-98	
	Amount	Percent (%)	Amount	Percent (%)
Revenue				
Sales-Labor	$46,292.16	55.04	$41,641.41	55.33
Sales-Parts	$37,806.89	44.96	$33,622.84	44.67
Total Revenue	**$84,099.05**	**100.00**	**$75,264.25**	**100.00**

You can see in Table 5-7 that labor sales and parts sales at Average Automotive have reflected a consistent 55/45 split for the two-year period reported and total sales have increased by $8,834.80. Sales have increased by almost 12 percent, while the percentage of parts and labor sales has remained constant, which should mean a significant increase in profits.

Labor Content Per Job

Labor content per job is one KPI that is often overlooked, and yet it can play a critical role in solving or at least ameliorating one of the most common complaints in the industry-the inability to schedule work in the shop efficiently. I stumbled across the idea of using labor content per job calculations to help better regulate the flow of work in our shop almost by accident while in the process of establishing a set of KPIs for our business. I was working on average invoice calculations when I began to wonder what role, if any, labor mix might play. After all, if the labor mix percentages are accurate, and they have to be, and your average invoice is actuarially sound, you should be able to apply the parts/labor ratios to your average invoice.

What do you gain by knowing what the average *parts dollars per job* or the average *labor dollars per job* per average invoice is? I wondered the same thing—until I actually did the math. Then it hit me! If you know what the average labor dollars per job is and you know what your labor rate is (preferably, your effective labor rate, which we will be discussing below), you can divide average labor dollars per job by your labor rate, and the result will be your labor content per job, as seen in Table 5-8.

TABLE 5-8   CALCULATING LABOR CONTENT PER JOB

LABOR CONTENT PER JOB		
Average Invoice		$168.35
Parts Sales in %	45	$75.76 Parts Dollars Per Job
Labor Sales in %	55	$92.59 Labor Dollars per Job
Labor Dollars per Job		$92.59
Divided by Labor Rate		$65.00
Labor Content per Job		1.4

The numbers are telling us that we can expect to spend an average of 1.4 hours on every vehicle that pulls down the driveway. Obviously, there won't be an hour and a half spent on every car. But the average is actuarially sound because all the numbers that led us to it, car count, average invoice, labor mix, etc., are actuarially sound.

Consequently, we can schedule about 5.3 cars for each technician capable of completing 8 hours worth of work in an eight-hour day, or close to 100 percent productivity. If we have ten cars on the lot, we know there will be 14 hours of work to be done regardless of what the vehicles might have been brought to the shop for.

In a minute we will be looking at service bay productivity. One of the concepts that will flow out of that discussion will be effective labor rate. Effective labor rate is another critical concept, and when we get there we will revisit the same calculation we just developed to see first-hand what dramatic difference productivity can make in your business.

Technician Efficiency

Generally, the issue of technician efficiency doesn't come up until after the discussion of service bay productivity has been addressed. This time, I'd like to do things a little differently and begin with technician efficiency instead of ending with it.

ef·fi·cien·cy, (ĭ-fĭsh'-ən-sē) *noun. Abbr.* eff **1.a.** The quality or property of being efficient. **b.** The degree to which this quality is exercised: *The program was implemented with great efficiency and speed.* **2. a.** The ratio of the effective or useful output to the total input in any system. **b.** The ratio of the energy delivered by a machine to the energy supplied for its operation. "An efficient person is competent or productive to a high degree upon some actual or imagined scale or model.".

We will define technician efficiency as the relationship between the time that has been sold or allotted for the completion of a specific service operation and the time it actually takes the technician to perform that operation.

Stated another way:

Hours Sold ÷ Actual Time = Technician Efficiency

The hours sold are the same as hours invoiced: they are the number of hours you multiply by your labor rate to determine how much you will charge the customer for the labor to perform the service, replace the part, or complete the maintenance operation. Actual time is the time it *actually* took the technician to complete the job, which is rarely the same as hours sold.

If the service writer at Average Automotive invoiced the customer $65.00 to remove and replace an alternator, which correlates to Average Automotive's $65.00 per hour labor rate, and the R&R actually took the technician the hour that was invoiced, technician efficiency for that particular job was 100 percent. If, on the other hand, it took the technician an hour and a half to remove and replace the alternator, technician efficiency would drop to 67 percent. With the proper tools, the proper training, the proper equipment, and the opportunity to succeed, such as the on-time arrival of the proper alternator, it is possible that the technician might have been able to remove and replace the alternator in less than one hour. In which case, technician efficiency would be greater than 100 percent. You can see these relationships in Table 5-9.

TABLE 5-9 TECHNICIAN EFFICIENCY EXAMPLES

TECHNICIAN EFFICIENCY				
Hours Sold + Actual Time = Technician Efficiency				
Hours Sold		Actual Time		Tech Efficiency (%)
50	+	40	=	125
50	+	50	=	100
40	+	50	=	80

The responsibility for technician efficiency falls at the feet of the technician, provided he is afforded the proper tools, training, technology, and the opportunity to get the job done. Productivity, as we will see in a minute, is a front counter (management) responsibility.

Service Bay Productivity

We will define service bay productivity as the relationship between the hours billed (or labor hours invoiced) and the labor hours available for sale (the hours the technicians are onsite and available to work).

Hours Billed ÷ Hours Available to Work = Productivity

Productivity has as much to do with the ability to sell the hours you have available as it does actually getting the work done. In fact, it may actually have more to do with selling those hours than anything else!

High Productivity = High Profits Low Productivity = Low Profits

It doesn't get any simpler than that.

If you have one technician who works 40 hours per week and you sell only 36 of those 40 available hours, you divide 36 by 40, and service bay productivity would equal 90 percent. If you sell 40 hours and the technician is able to complete the 40 hours of work you sold in the 40 hours he was there, service bay productivity would be 100 percent. And, finally, if you sold 50 hours of labor that your technician was able to complete in the 40 hours he was onsite and working, service bay productivity would soar to 125 percent (see Table 5-10).

TABLE 5-10 TECHNICIAN PRODUCTIVITY EXAMPLES

SERVICE BAY PRODUCTIVITY				
Hrs Inv	+	Hrs Avail	=	Productivity (%)
36	+	40	=	90
40	+	40	=	100
50	+	40	=	125

Unfortunately, statistics suggest that service bay productivity numbers leave a lot to be desired, especially for the independent segment of the service industry. And the tragedy there is the profound impact productivity has on every aspect of a service business. The example above is representative of one technician. What happens when there are three technicians working in the shop? They won't share the same numbers, especially if they are segmented—by that I mean specializing in a certain area of repair. For instance, you might have one technician doing high-tech, drivability diagnosis and repair-focusing on intermittent and erratic electronic problems. Another technician might be doing nothing but brake and front-end work, and still another might spend most of his time focusing on light and heavy line. Their productivity numbers will reveal a significant spread in most shops around the country, with the highest productivity numbers coming from the brake and front-end specialist, the lowest numbers coming from the drivability specialist and the hours closest to 1:1 (one hour billed for every hour worked) coming from the technician doing the light and heavy line work (see Table 5-11).

TABLE 5-11   AVERAGE PRODUCTIVITY IN A THREE-TECHNICIAN SHOP

AVERAGE PRODUCTIVITY: 3-TECHNICIAN SHOP			
	Sold Hours	Avail Hours	Productivity (%)
Driveability Tech	26	40	65
Line Tech	34	40	85
Brake & Front End	44	40	110
	104	120	87

Average service bay productivity of 87 percent would actually be something you could be proud of, especially when most of your peers are posting numbers below 70 percent. But that isn't the issue I would like to focus on for the moment. I would like to turn your attention to what the numbers mean to you and possibly your shop. Service Bay Productivity directly impacts a key performance indicator—I refer to it as Effective Labor Rate, or the labor rate you have posted on the wall factored by Service Bay Productivity. What does that mean? Let's work with the same $65.00 per hour labor rate Average Automotive has been posting throughout this discussion. In most cases, that's the labor rate Average Auto's owner *thinks* he's getting for every hour he bills. It isn't! Because his shop is running at an average service bay productivity of 87 percent, he is actually only receiving 87 percent of $65.00 or $56.55. Labor Rate x Average Service Bay Productivity = Effective Labor Rate

If Average Automotive was realizing an effective labor rate closer to the averages I've seen over the years or the numbers reflected in most productivity surveys, the effective labor rate would be closer to $35.75. In another guide in this series dealing with achieving operational excellence, we'll take a long, hard look at all the factors that can impact and influence service bay productivity. But for now, I'd like to limit the discussion to effective labor rate and labor content per job.

When we looked at labor content per job earlier in this chapter, we calculated it by taking the dollar amount of our average invoice and multiplying that by the percentage of labor mix found in our financial statement or calculated through our computerized automotive shop management software. The result was the labor content per job expressed in dollars. We then took that number and divided it by our labor rate to give us our labor content per job expressed in hours. But we just saw that that would only—*could only*—be accurate if our productivity was 100 percent! In most shops across the country, we're not receiving one hour's worth of revenue for every hour we're there because we are either not selling every hour we have to sell or our technicians are unable to produce at 100 percent technician efficiency. Consequently, in order for our labor content per job (expressed in hours) to be accurate, it must be factored for service bay productivity.

Look at what happens in Table 5-12 when we do that.

TABLE 5-12   THE IMPACT OF 87 PERCENT SERVICE BAY PRODUCTIVITY ON LABOR CONTENT

LABOR CONTENT PER JOB		
Average Invoice		$168.35
Parts Sales in %	45	$75.76 Parts Dollars Per Job
Labor Sales in %	55	$92.59 Labor Dollars per Job
Labor Dollars per Job		$92.59
Divided by Effective Labor Rate ($65.00 X 87% = $56.55)		$56.55
Labor Content per Job		1.6

Our labor content per job went from 1.4 hours to 1.6 hours. What does that mean? It means that it will take us 0.2 hour longer to finish each car we see in the shop, or one less car per day-$168.35 less per day in revenue. If we modify the labor content per job to bring it more in line with national averages, adjusting the effective labor rate to reflect average productivity of 55 percent, the numbers get even worse, as seen in Table 5-13.

The labor content per job goes up, but the size of the average invoice does not! It now takes us 2.6 hours to complete $92.59 worth of labor billing, and we can only work on three cars a day effectively.

High productivity is critical to your success as a shop owner or manager. It impacts everything from bottom line profits to scheduling, from staffing to just about everything else that happens in the shop. When we begin to discuss operational excellence, we'll take this critical KPI apart, and then we'll take it apart again!

TABLE 5-13   THE IMPACT OF 55 PERCENT SERVICE BAY PRODUCTIVITY ON LABOR CONTENT

LABOR CONTENT PER JOB		
Average Invoice		$168.35
Parts Sales in %	45	$75.76 Parts Dollars Per Job
Labor Sales in %	55	$92.59 Labor Dollars per Job
Labor Dollars per Job		$92.59
Divided by Effective Labor Rate ($65.00 X 55% = $35.75)		$35.75
Labor Content per Job		2.6

Total Sales per Bay

The final two KPIs we will be focusing on here are total sales per bay and total sales per tech. Total sales per bay is relatively simple to calculate. You just take the total sales for the period you are focusing on and divide it by the number of bays you have. If Average Automotive had three bays and sales of $33,333 per month, total sales per bay for that month would be $11,111.

That is the actual number. But, what is the target?

The number I have heard thrown around most is $20,000 to $25,000 per bay per month, and it is calculated on the following projection, assuming one technician in each service bay (Table 5-14).

TABLE 5-14   TOTAL SALES PER BAY CALCULATION

TOTAL SALES PER BAY	
Techs per Bay	1
Hrs per Day	8
Days per Month	22
Tech Hrs per Month	176
X Labor Rate	$65
Total Poss Labor	$11,440.00
Plus Est pts Sales	$11,440.00
Total Sales per Bay	$22,880.00

Unfortunately, this number can be misleading. Not every shop has one technician for each service bay, and consequently, since the revenue generator is the technician and not the service bay, the calculations fall apart. In other cases, one or more of the existing service bays may be unavailable for any one of a number of reasons, including storage for unused or under-utilized equipment, the owner's personal boat, race car, or truck, or just plain clutter.

Aside from that, few shops run at 100 percent productivity (meaning eight hours of billing per day, which is what the calculation above would suggest). In fact, looking at the actual performance of Average Automotive versus the target projection, what do you see? Actual performance of about half the projection—a target of $68,640 (3 techs x $22,880) and actual sales of $33,333, or just under 50 percent productivity.

In this case, a more accurate KPI might by total sales per tech.

Total Sales per Tech

Total sales per tech are calculated the same way as total sales per bay, with one exception—the number of service bays is irrelevant. However, one thing might become obvious if you play with the same numbers we used above. Average Automotive has three technicians and total sales of $33,333 per month. Our calculations would suggest that if Average Automotive was working at 100 percent productivity, those sales should be closer to $100,000 per month, not one-third of that!

SUMMARY

The key to KPIs is first understanding what the numbers are trying to tell us, and then using what those numbers are trying to tell us to drive up performance. The numbers are only numbers if we don't use them to create and sustain positive movement and continuous improvement.

The key to success in our industry is to increase technician efficiency, raise service bay productivity, increase average invoice, develop a higher percentage of labor sales to total sales while maintaining or increasing parts sales and growing labor content per job. If we can do all that, we will begin to see our bottom line float up to where it really belongs!

How will you do that? Write your thoughts in the space on the next page, then track your actions and their results.

Thoughts

Actions

Results

CHAPTER 6

Business Planning

INTRODUCTION

This chapter introduces you to the idea of a business plan by asking and then answering the following fifteen critical questions about both your business and the industry you serve:

- What business are we in?

- Who are our customers?

- What do they want, need and expect from us?

- How do we make our products and service available?

- Who is our competition?

- What advantages do we have over our competition?

- What advantages do our competitors have over us?

- What do our competitors do that we don't do?

- What do we do that our competitors can't or won't do?

- Who are the owners of this business and what are their roles and strengths?

- Who are the managers of this business and what are their responsibilities?

- What financial information do we have available and when?

- What financial information do we need?

- What changes will take place in our type of business over the next two or three years?

- What are our company's current problems or desired goals?

It will then help you put the answers to work for you.

BUSINESS PLANNING

Question: When do most business owners go to the bank or to a private lender for financial assistance?

Answer: When they are desperate and there is no alternative!

Question: What kind of information do they have available to help facilitate this request for financial aid?

Answer: Most of the time we'll never know because they don't know and consequently go to the lending institution empty handed and unprepared!

Question: What kind of information do they need?

Answer: Whatever it would take to make a compelling argument on their behalf, like: How much money will be needed? What the money will be used for? How it will be paid back? How long it will realistically take to repay the loan? What kind of return this investment will bring, both for the service provider and for the lender or investor?

Question: How can you tell who has a business plan and who does not?

Answer: By who gets the money!

You may think some of these questions and answers are a bit tongue in cheek, but they're not. Not really. Most business owners are rarely prepared to speak to their banker in a language the banker understands and appreciates, let alone someone they don't know or someone unfamiliar with their business or our industry. Consequently, I think it would be wise to spend a few minutes talking about what is generically referred to in business as a business plan. In *From Intent to Implementation*, another guide in this series, we spend more time on this subject. But we'll be approaching the business plan more from a financial perspective than from a philosophical or practical one. For now, however, we will be concentrating on building a simple, yet practical plan for your business—a general kind of road map to keep you moving in the right direction.

There are a number of different ways we can approach business planning. They range from the most comprehensive and detailed plans, to plans that are simple, general, and yet still practical. You can find volumes written on the subject, complete with countless templates that require just plugging in your numbers. However, what we will be discussing here are a number of different things that when stitched together make sense in our environment. To a large degree, that means they have to be compatible with everything else we've discussed so far, and they have to be practical and easy as well. But most of all, they must prove effective.

THE FIFTEEN-QUESTION BUSINESS PLAN

I suggest starting with fifteen sheets of paper, one for each of the fifteen questions I am about to share with you. And, since this is a plan for your business, I suggest that you have one set of these questions for each of the individuals working inside or outside your business with a stake in your success. Place a question at the top of each sheet of paper and then create a complete set for everyone involved. The questions I will submit here can be changed or modified to suit your particular needs, but I suggest starting the exercise with these questions first, just to get an idea of where the answers will take us. Answering these fifteen questions will afford you the foundation you need to start building a plan which not only makes sense to you, but which will also help an interested third party understand who you are, what you are all about, and where you are taking your business.

What Business Are We In?

My hope is that your answer will vary depending upon which of the other guides in this series you've read first. Realistically, most automotive service professionals would say they are a part of the automotive service industry; in the business of servicing, maintaining, and repairing passenger cars and trucks. While that is certainly an accurate answer, I'm not sure it is an adequate one. In today's highly chaotic and competitive environment, an environment in which consumers have never been more knowledgeable and less forgiving, servicing, maintaining, and repairing passenger cars and trucks may be enough to get by, but it just isn't going to be enough to insure your success anymore. Fixing the car right the first time is just what it takes to get in the game, or to stay there once you are in! It is nothing more or less than the price of admission.

It will take meticulous attention to every detail involved in meeting or exceeding the customer's wants, needs, and expectations consistently and without fail each and every time you are asked to perform. It will take having the vehicle ready when promised—not every once in a while, not most of the time, but just about every time. It will take being there when the customer needs you to be there, and it will take an unparalleled quality of both parts and service.

So the answer to the first question would certainly have to include the fact that you are in the automotive service industry, servicing, maintaining, and repairing your customer's cars and trucks. But it would also have to include a statement of purpose consistent with meeting or exceeding customer wants, needs, and expectations as well—perhaps something like this:

> *Schneider's Automotive Repair is in the business of insuring the personal freedom and mobility of our customers through the care, maintenance, and repair of their passenger cars and trucks. We do this by understanding and anticipating their wants, needs, and expectations almost before they do and then by meeting or exceeding those wants, needs, and expectations each and every time service is delivered, by providing them with the needed work, done well, finished on time, and at a fair price with all work guaranteed.*

Who Are Our Customers?

Once again, the answer to this second question might be different depending upon where you find yourself in this series or on this journey. If you understand how important it is to choose the right target market for your products and services, you have already come to realize that you can't work for everyone, because not everyone will appreciate the level or quality of the products and services you provide. You really have to take the time to determine the level of quality and service you will demand from yourself and from your associates and then profile the kind of customer who will respond. You must either do that or survey the demographics of the neighborhood you will be serving, and then profile a target market within that neighborhood, focusing on the kinds of products and services they both want and need. Either way, you will have to describe your target customers (more than one target) in great detail.

One of our targets is: *Single women, heads of households, gross income of $35,000 or more.* Our description is really quite a bit more detailed than that and includes a number of additional qualifiers like the makes and models of the cars, trucks, and SUVs that they would be driving and that we prefer to work on. Another target might be: *Boomer geezers—a term that is becoming increasingly popular, used to describe the 76,000,000+ Americans who are now 55 years old and older, with three drivers or more in the family and a gross family income of $100,000 or more.* Again, this particular target can be defined in significantly greater detail. However, either example should give you some idea of how you can qualify and then quantify the kind of people you would like to invite to experience the products and services your business has to offer. But in the end, it would be wise to try to find the kind of targets that fit the profiles of those customers who make you smile every time they pull into the driveway. That way you'll always feel good about the commitment and effort required to deliver world class service to all your customers all the time.

What Do They Want, Need, and Expect from Us?

In any discussion of customer wants, needs, and expectations we cannot escape the obvious, which is—*fixed right the first time*! But in every customer satisfaction survey we've seen over the past few years, that is not the number one customer concern. The number one issue for the customer is—*ready when promised*! Having the vehicle available when the service facility said the job would be completed is critical to lasting customer satisfaction. Having a plan that ignores something that critical would be incomplete at best, dangerous and misleading at worst.

Time is perhaps the most critical commodity of our era. Everyone wants more, no one has enough. The demands of work, friends, and family have each of us stretched to the limit, and our failure to be sensitive to that reality leaves the customer no choice but to think we don't care.

So in answering a question like the one above, I would start with the consumer's need to have the vehicle *ready when promised*, then go on to *fixed right the first time*. I would then follow those two with explanations of meeting customer wants, needs, and expectations of convenience, quality, accessibility, and response-ability—the ability to respond quickly and completely to the customer's demands.

How Do We Make Our Products and Services Available?

This question goes to the heart of meeting or exceeding customer wants, needs, and expectations by understanding when and how they want to access our products and services. It has to do with staffing and hours of operation, with before- or after-hours vehicle pick-up or delivery, and it has to do with how we package those services.

Do you plan on being open in the evenings, on Saturdays, or even on Sundays? If it was up to most vehicle owners, you or someone like you would be at the shop 24/7/365. But barring that, when *will* you be open—early enough to catch your customers before they have to leave for the office? And when will you close—late enough to give them a chance to stop by, pick up the vehicle, and pay you?

These are all questions that can and should be addressed in the kind of plan we are discussing here.

Who Is Our Competition?

In the midst of a heated discussion centered on neighborhood competition, a good friend of mine taught me an invaluable lesson. He was the consummate businessman, and as I complained endlessly about what my competitors were either doing or not doing, he quietly said, "I have no competition!"

At first, I was really amazed. I know his business and I know how and where he goes to market, and believe me, it is not without its share of competition. But when I pressed him for an explanation, his meaning became clear.

> *I'm the competition! I'm the one everyone has to watch out for! I'm the innovator! I'm the one committed to understanding my customer's wants, needs, and expectations better than anyone else in my market. I'm the leader. They're the followers. By the time they have imitated something I did yesterday, I've already evaluated whether or not it was working and gone on to something else. I'm always moving forward, while they're always trying to catch up! Since I'm out in front by myself, I have no competition!*

That one statement changed my business life, and now I can honestly say that I try to run my business as if I have no competition as well. However, for the sake of a business plan, identifying groups of competitors is not only prudent, it's vital. No one will take your plan seriously if it does not include at least a list of your competitors, a description of who they are and an explanation of what they do. My recommendation is to place the other service facilities in your market area in tiers: first, second, third, etc. First-tier competitors would be those service facilities that provide their customers with the same quality of service you do. The easiest way to identify such competitors would be to ask yourself who you would send one of your favorite customers to if you were too busy to work on the vehicle yourself and for some reason it absolutely had to be worked on by someone. My guess is that there are at least two or three service facilities in your general area that would fit that criterion. This first tier could also include a number of factory service outlets.

The second tier would be comprised of specialty shops, automotive franchises, retailers, and mass marketers. The third tier could be the quick service facilities, and so on.

Describe these outlets in as much detail as you are able, as completely and accurately as you can.

What Advantages Do We Have Over Our Competition?

This is a really important question to answer clearly, honestly, and accurately. It will help you understand exactly what your service capabilities and capacities are when compared with other service providers in your market area. Answers could range from location to technical capability, from hours of operation to number of service bays. But whatever the answer, it should call for a fairly extensive assessment and consequently an honest evaluation of just where your products and services are superior and the competitive advantage that results from those superiorities.

What Advantages Do Our Competitors Have Over Us?

Just like the answers to the question above, the answers to this question require careful assessment and evaluation. However, unlike the question above, this question is inherently more difficult to answer honestly because it forces you to confront your own weaknesses and/or failures. It forces you to look at where your competition is really capable of using adjectives like *more* and *better*. But it is through this process that you will uncover the areas you will need to confront in order to achieve a leadership position in the marketplace.

What Do Our Competitors Do That We Don't Do?

Once again, we need to look at things that just might make us a little uncomfortable. Like whether or not our competitors are open when we are not. Or whether they have made an investment in training, equipment, or technology that allows them to perform in areas that we are not able to address. It almost demands that you develop a service menu for your facility and then a comparative list for your competitors. By the process of evaluation, you will not only have the answers to this question-you will have the answers to the next question as well.

What Do We Do That Our Competitors Don't Do?

The answer to this question will materialize as you answer the question above. In essence, it is the things we are capable of doing well, that our competitors can't or won't do. The answer could be a specific service or a whole area of services. Whatever it is, it will point the way toward a competitive advantage, a competitive edge.

Who Are the Owners of This Business, and What Are Their Roles and Strengths?

In too many instances, this may be the first time this question has ever been asked and answered. A clear description of the owner's responsibilities, roles, and strengths is critical for success—especially when there is someone in a secondary management position such as a general manager or a service manager. Without such a description, responsibilities become cloudy, and things, sometimes very important things, fall through the cracks.

Perhaps the best example of all this is what we experienced in our own business. There are four Schneiders at Schneider's Automotive. It wasn't that long ago that we were stepping all over each other's responsibilities, doing each other's jobs. Ultimately, the chaos created by never knowing if or when something got done, or if more than one of us did the same thing, became so counter-productive we finally had to split things up into specific areas of responsibility with formal job descriptions. As uncomfortable as this can be, it is sometimes the only way to insure that everything that needs to be accomplished is accomplished.

Who Are the Managers of This Business and What Are Their Responsibilities?

The same person who taught me about having no competition taught me another valuable lesson as well. He taught me that you *hire your weaknesses*! The concept is simple. We all like to be around people like us! However, in business, you don't really need to hire someone to do something you are really good at doing when you could or should be doing it yourself, unless you no longer want to do it. It is more important, critical in fact, to hire individuals capable of doing well what you are not good at doing at all—the things you no longer want to do or were never very good at. If you are not a detail-oriented person, you need to hire someone who is meticulous in everything they do, someone obsessed with detail. If you're the one who is detailed and you find yourself reluctant to get close to your customers, you need to hire a people person to do that job. The same holds true for organizational skills and technical competencies.

Of course, in order to do that, you have to understand and know your own strengths and weaknesses and staff your technical and management positions with those individuals who are strong where you are weak.

What Financial Information Do We Have Available and When?

I have actually been in shops when the financial statements arrived and watched as the owner/principal opened the manila envelope, removed the financial statements, looked to insure the number at the bottom of the page wasn't in brackets, and then slid the financials back into the envelope never to be looked at again! That is not financial analysis. It's financial suicide!

This information is critical, and because financial information is really historical in nature—it is derived from information concerning events that have passed—the sooner you get that information, the better. In other words, you need everything you can get and you need it as fast as you can get it. Time really is of the essence in every sense of the expression. It is equally important to have that information adequately represent the actual performance of your business.

What Financial Information Do We Need?

Certainly we need a full array of financial statements, including, but not limited to, a current income statement or statement of accounts, a current balance sheet, a cash flow analysis, and a break-even report. You need an accurate and current listing of your accounts receivables, money that is owed you. And you need an accurate and current listing of your accounts payable, money that you owe. You also need a current and accurate inventory report. It's always a good idea to know roughly when you will actually be working for yourself, so a break-even analysis is something I would suggest as necessary.

In the end, however, it's really no different from what you are willing to do for your clients or what they could and should expect from you. It is their vehicles and their money, so it's really their responsibility to find a service provider able and willing to explain their automotive service needs in terms they can understand so they can make the right decisions and take the right action. And it's really your business and your money, and consequently your responsibility, to find an accountant or a bookkeeper that is willing to explain everything you need to know in terms you can understand.

What Changes Will Take Place in Our Type of Business Over the Next Two or Three Years?

It would be wonderful if we could answer a question like this, but in all honesty we probably can't. However, we *can* project what we know about both the past and the present into the future. For instance, we know that the explosion of technology we have been confronted with over the past thirty years is not going to stop. We know we will be looking at hybrid vehicles, 42-volt electrical systems, and ultimately fuel cells. We can be pretty sure that the shortage of trained and qualified technicians we've been experiencing isn't going to somehow fix itself. And we know the competition for automotive service dollars will become more intense as well.

If we know all that, we can begin to think about how we will respond to those challenges, and that is what the planning process is all about.

What Are Our Company's Current Problems and/or Desired Goals?

By taking a long, hard look at our business, its capacity and capability to deliver world class service consistently, we have gone one step beyond the majority of other service providers working in our industry. By identifying our problems and listing our opportunities, by itemizing the threats and considering both our strengths and our weaknesses, we have demonstrated that we are aware of the environment and conditions in which we perform and have both the desire and the ability to succeed. Those are the kinds of things financial professionals look for when analyzing a business or considering a request for financing. They are likely to have infinitely more faith in someone with a plan than in someone willing to go where the tide takes them, someone more likely to *react* than *act*.

SPECIFIC ELEMENTS OF A PLAN

A formal business plan should include both a statement of the company's vision as well as a mission statement. As we have discussed in a number of the other guides in this series, that mission statement should clearly describe what you perceive to be the company's long-term mission.

It should also include professional biographies for each of the individuals involved in managing the business, as well as biographies for key employees and the owner or owners. These biographies should encompass where they have been and what they have done while involved in the industry they serve, their major strengths and accomplishments, and the number of years they have been involved. It wouldn't be a bad idea to include an explanation of the automotive service aftermarket: its history, the industry as it exists today, and projections for the future. Despite what we would like to believe, not everyone is familiar with our industry. Consequently, this analysis should include a comprehensive overview of the marketplace, a description of who the key players are and how they fall into your competitive matrix, and an explanation of how and why you will not only survive but succeed. In doing that, your analysis should address the company's strengths, weaknesses, market opportunities, and threats.

It would be wise to emphasize your understanding of the marketplace by identifying certain consumer problems along with an explanation of how you plan to exploit the opportunities created by those problems. All of this should be wrapped around your philosophy of doing business, your business concept. Identify the various tiers of competitors as described earlier, and explain how and why you feel you have a competitive advantage. Enumerate your company's goals, objectives, and tactics—what you are going to do in order to succeed and when you're going to do it. These can and should be broken down into long-term (3-5 years) and short-term (within 1 year) goals and the specific objectives associated with each. State your specific goals for customer share and market share and how you intend to achieve them. Identify your specific financial goals, particularly in reference to your KPIs, and show how achieving those goals will greatly enhance your overall financial picture. Discuss what it's going to take to make all this happen—the resources, technology, and staffing necessary. And then talk about risks and rewards—what they are and how they will be addressed.

Recognize the fact that by taking the time to do any or all of the above you are a part of a very small minority of small business owners who have realized that information is power—the power of knowledge, the power of confidence. And know that armed with that knowledge, you are in a much better position to guide your business to success.

SUMMARY

If you fail to plan, you plan to fail! It would be hard to argue against that statement, and yet many of us still fail to invest the time, energy, and effort required for effective business planning. Do we know our product? Do we know our customers? Are we aware of the competition and the marketplace and where we fit in? Have we thought about the future and what it has in store for us and those close to us? Could we change that future? Could we make it different somehow if we took a proactive role in managing it?

Can you answer the questions in the fifteen-question business plan? Can you think of additional questions that might help you prepare for the future?

Think about what you can do to change the reality you face today and the kind of future you would like to face tomorrow, and then write your thoughts in the area below.

Thoughts

Actions

Results

CHAPTER 7

Conclusion

While the title *Managing Dollars With Sense* may seem like a superficial attempt at cute through an obvious play on words, it's really not. The only way to manage dollars is with sense, good sense, common sense—the kind of common sense we sometimes lose when we are too deeply involved in *now* and not involved enough in *later*.

We could just as easily present the title of this volume as *Managing Dollars with Cents*. It may not have been as clever, but it may have been just as accurate. After all, you can't really manage the dollars unless you pay attention to where the cents are going. You can't really manage anything if you don't understand or appreciate exactly what it is you are looking at. That's why you will see certain key performance indicators reintroduced, discussed, and then reinforced throughout this series, and that's why I began this particular volume with a discussion of exactly where the numbers come from and what they mean.

For years I have insisted that learning how to perform the best brake service possible is irrelevant if you can't make enough money to allow you to perform that service the way it needs to be performed. It may seem like heresy to suggest that it doesn't really matter how well you can fix cars anymore, but it doesn't if you don't know how much it is actually costing you to fix them. Unfortunately, you won't have the opportunity to fix the cars or serve the customers very well or for very long if you can't pay the rent, your suppliers, your technicians, or yourself! That's why the second chapter was all about understanding your financials and the critical ratios that tell you and the financial professionals you will be forced to interact with just how well you are doing.

I have tried to make *Managing Dollars With Sense* as comprehensive as possible. To me, that means including the basics such as margin and markup, as well as some of the more sophisticated concepts like reverse budgeting, avoiding Oreo pricing, inventory, what I consider to be the ten most critical KPIs, and CODB. In *Managing Dollars With Sense* you now have one more tool in your tool box—one more way to move closer to your dreams.

You don't have to Oreo price. You don't have to wonder how you're doing or even how or where to start improving your situation. It's all here, or at least as much as you need to change your current reality—as much as you need to move the needle in the direction of success.

I have tried to provide you with the technology necessary to make dealing with now a bit less stressful and *managing dollars with sense*, a little bit easier.

I've given you a way—a simple, easy way—to build an elementary business plan and the business knowledge to help take that plan *from intent to implementation*. But in the end, it's all up to you. It always has been. You will ultimately decide whether the tool comes out of the toolbox or whether it stays in the bottom drawer. You will either continue to do things just as you have always done them, or you will dare to see things as they could be, and then act upon those dreams.

If you always do what you've always done, you will always get what you've always got!

The question you have to answer is whether or not what you've got is good enough for today, *and* good enough for tomorrow.

MORE TO COME

This series, *Automotive Service Management*, is divided into three general areas of concentration and eight guides. As we make our way through these eight guides, we will be looking at virtually everything involved in successfully operating a contemporary automotive service facility in today's dynamic and challenging marketplace. That means we will be looking at such things as hiring practices, productivity and service management, the cost of doing business (factored for service bay productivity), cost/value ratios, value-added service, and hazardous materials communication. As well as exploring our relationship with the consumer, our suppliers, our employees, and our industry, we will also be looking at promotions, advertising, sales and marketing.

Our goal is to create the most comprehensive resource for automotive shop management available anywhere—a complete and exhaustive look at all the *hard* and *soft* skills required to manage any business, as well as the tools required to run an efficient, effective, customer-responsive automotive service facility.

The series was not conceived or formulated in a vacuum. It did not originate in theory first, and then find its way into reality. It is the result of a lifetime spent in the automotive industry, backed up by four generations of genetic memory. I invite you to take whatever you can from these pages. Take the key performance indicators you find here and create a snapshot of your business as it is at the beginning of this journey. Then see how integrating what you find within the pages of these manuscripts changes those numbers, how it impacts the bottom line—your bottom line. And then let us know how you're doing!

In addition to this text, *Managing Dollars With Sense*, there are seven other titles for you to read, enjoy, and work through, including: *Total Customer Relationship Management, From Intent to Implementation, Operational Excellence, Building A Team, The High Performance Shop, Safety Communications,* and *Operations Management.*

If you have any suggestions for subsequent subjects or ways we can improve what we have already presented, let me know. My goal is provide you with answers to your questions—answers I had to search for or create myself. My goal is to insure that you don't find yourself struggling the way I had to. My goal is to see that you are richly rewarded for your efforts and your ability. My goal is to see you succeed where others have failed.

So read this guide one section at a time or all at once. See what you like. See what you aren't all that comfortable with. *Then read it again!* Find those principles that make sense to you. Create a plan to integrate them into your business. *Then read it again!* Make the changes required to take your business to the next level. Monitor your progress. Document everything. *Then read it again!*

If it works for you, let us know. If you can find ways to help us make it work for you more effectively, let us know that too!

APPENDIX a

Ratios

LIQUIDITY RATIOS

Quick Ratio

Quick Assets (Cash + Accounts Receivable)	
Cash	$11,247.06
+ Accounts Receivable	+ $6,078.53
= Quick Assets	= $17,325.59
÷ Current Liabilities	÷ $19,880.74
Quick Ratio	87.15%

Current Ratio

Current Assets	$63,407.48
÷ Current Liabilities	÷ $19,880.74
Current Ratio	3.1:1 (318.94%)

Working Capital

Current Assets	$63,407.48
- Current Liabilities	- $19,880.74
Working Capital	43,526.74

Receivables to Payables

Receivables	$6,078.53
÷ Payables	÷ $9,840.94
Receivables to Payables	**.6:1 (61.77%)**

FINANCIAL PERFORMANCE RATIOS

Gross Profit Percentage

Gross Profit	$36,346.28
÷ Net Sales	÷ $67,660.13
Gross Profit Percentage	**53.72%**

Return On Equity

Net Income	$9,539.84
÷ Stockholder's Equity	÷ $20,398.75
Return On Equity	**46.77%**

Return On Assets

Net Income	$9,539.84
÷ Total Assets	÷ $96,252.79
Return On Assets	**9.91%**

Debt Ratio

Total Liabilities	$75,854.04
÷ Total Assets	÷ $96,252.79
Debt Ratio	**.78:1 (78.81%)**

Equity to Assets Ratio

Total Stockholder's Equity	$20,398.75
÷ Total Assets	÷ $96,252.79
Equity Ratio	**21.19%**

Operating Profit Margin

Net Profit	$9,539.84
÷ Total Sales	÷ $67,660.13
Operating Profit Margin	**14.10%**

Debt to Equity

Total Liabilities	$75,854.04
÷ Net Worth	÷ $20,398.75
Debt to Equity	**3.72:1 (3.72%)**

Return On Investment

Net Profit	$9,539.84
÷ Net Worth	÷ $20,398.75
Return On Investment	**46.77%**

GLOSSARY

Account

An individual record or a niche used to accumulate similar information relating to specific stuff the business sells, commonly referred to as assets; stuff the business owes or pays, commonly known as liabilities; and stuff the business owns, commonly called capital.

Accounts Payable

This is an amount of money that you owe a supplier or vendor for goods or services performed on credit. These accounts are generally classified as current liabilities because they are almost always due and payable within twelve months (30 days in most cases).

Accounts Receivable

This is an amount of money owed the business by a customer, an individual, or a fleet account, for services rendered, work performed, or parts purchased.

Accrual

A method of accounting that records income as it is incurred but not yet received and expenses as they are incurred but not yet paid.

Asset

Anything you own that has monetary value.

Balance Sheet

A document that reports the financial position of a business entity at one specific moment in time. It is a detailed presentation of the assets, liabilities, and owner's equity in that business. It is an expression of the accounting formula in which the total value of the assets is equal to the total of all the business's liabilities added to the owner's equity or capital;

Assets = Liabilities + Owner's Equity

Capital

The ownership of the assets of a business by the proprietor(s).

Cash

An asset consisting of coins, bills, money orders, checks, credit card vouchers, certificates of deposit, or treasury bills.

Cash Flow

The cash flowing in and out of a business through receipts and payments.

Cost of Goods Sold

The cost of goods sold in our industry is generally calculated by taking your beginning inventory, adding to it your net purchases for that accounting period, and then subtracting your ending inventory. A number of shops across the country add to the cost of goods sold the expense of shop labor. This equation expressed on a profit and loss statement generally looks like this:

Net Sales - Cost of Goods Sold = Net Profit

Chart of Accounts

A chart of accounts is more or less a table of contents, a listing of all the accounts that make up a ledger and their account numbers.

Current Assets

These are assets such as cash, marketable securities, accounts receivable, inventory, and pre-paid expenses which it is reasonable to believe will be converted into cash by sale, or used up, within a twelve-month period.

Current Liabilities

These are debts that must be recognized and paid within one calendar year (or the current accounting period, whichever is longer).

Current Ratio

A means by which a business can determine its ability to meet its financial demands (pay its bills) when they are due. This ratio is one of the two key tests to determine just how liquid your business assets really are. The formula for current ratio is:

Current Ratio = Current assets ÷ Current Liabilities

Depreciation

An accounting term used to designate the systematic and rational allocation of the cost of an asset over its useful life.

Expenses

The costs that must be incurred in order for a business to generate revenue. These may include wages, salaries, advertising, taxes, insurance, depreciation, interest, utilities, rents, etc.

Fixed Assets

Fixed assets are generally considered to be assets with an expected useful life of one year or more and may include land, building, capital equipment, etc.

Gross Profit

The revenues generated by sales, less the cost of goods sold.

Income Statement

A financial document that presents revenue and expenses and the net income or loss for a specific period of time.

Inventory

The parts and accessories that you stock on your premises for sale to your clients. On your balance sheet, inventory is included under the heading of current assets.

Inventory Turnover

The number of times the gross dollar amount of your inventory is sold and restocked over a given period of time, usually one year. The formula for inventory turnover is:

Inventory Turnover = Annual Cost of Inventory ÷ Average Monthly Inventory

Liquidity

The ability of a business to meet its daily or monthly cash demands from its current assets. Two key ratios to determine liquidity are the current ratio and the quick ratio.

Long-Term Liabilities

An obligation incurred by a business that is not expected to mature within one calendar year. A mortgage is a good example of a long term liability.

Net Income

Synonymous with net profit—most commonly referred to as the *bottom line* in business. Net income is what is left of your revenues after all costs, expenses, and tax obligations have been met.

Net Profit

The same as net income.

Net Worth

On your balance sheet, a calculation of your equity as an owner or a stockholder. The formula is:

Net Worth = Assets - Liabilities

Operating Expenses

These are the various expenses incurred by a business in its daily operation. They include wages, taxes, uniforms, laundry, insurance, utilities, interest, depreciation, advertising, and a host of other expenses.

Prepaid Expenses

These expenses are an asset account because even though they are an expense, they are paid before they are due. At such time as the value of that prepaid asset has been used up, an adjusting entry is made that will convert this prepaid expense (asset) to an actual expense. A good example of this kind of prepayment is an insurance policy purchased and paid for at the beginning of the year, or prepaid quarterly taxes.

Profit & Loss Statement

Often referred to as an income statement. This is a financial document summarizing the income and expenses incurred by a business over a set period of time.

Profit

The revenue remaining after expenses.

Profitability

The return on investment a business provides the owners or stockholders.

Revenue

The sale of a product or service for which assets, such as cash or accounts receivable, are received, ultimately having an effect on owner's equity.

Quick Ratio

A calculation which measures the ability of a business to pay its bills when they come due. The quick ratio is very much like the current ratio-the major difference is that the quick ratio does not include inventory, as inventory is not easily or quickly converted into cash. The formula for quick ratio is:

Quick Ratio = (Current Assets - Inventory - Prepaid Expenses) ÷ Current Liabilities

Return on Investment

A calculation measuring the earnings that have been generated utilizing all the capital invested by all the owners of a business. This calculation may also be called a return on net worth or a return on equity. The formula is:

Return on Investment = Annual Net Profit after tax ÷ Net Investment (beginning of the year)

Short Term Debt

Financial obligations due and payable within twelve months.

Working Capital

The money required by a business to cover day-to-day expenses. The formula is:

Working Capital = Current Assets - Current Liabilities

This formula is derived from information normally found on a balance sheet.

REFERENCES

The American Heritage Dictionary of the English Language, 3rd ed., Boston, MA: Houghton Mifflin Company, 1992.

Pierce, Edward. *Accounting The Easy Way*, 2nd ed., New York: Barron's Educational Series, Inc. 1989

Ed Chadroff, CPA
Segal, Chadroff & Wolff
280 East Thousand Oaks Blvd.
Suite B
Thousand Oaks, California 91360
(805) 497-2291